Sustainable Livestock Management
for Poverty Alleviation and Food Security

Sustainable Livestock Management for Poverty Alleviation and Food Security

Katrien E. van't Hooft, DVM

Endogenous Livestock Development Network, TradiNova Livestock

Terry S. Wollen, DVM

Heifer International

Dilip P. Bhandari, BVSc & AH, MVM

Heifer International

www.cabi.org

CABI is a trading name of CAB International

CABI	CABI
Nosworthy Way	875 Massachusetts Avenue
Wallingford	7th Floor
Oxfordshire OX10 8DE	Cambridge, MA 02139
UK	USA
Tel: +44 (0)1491 832111	Tel: +1 617 395 4056
Fax: +44 (0)1491 833508	Fax: +1 617 354 6875
E-mail: cabi@cabi.org	E-mail: cabi-nao@cabi.org
Website: www.cabi.org	

A catalogue record for this book is available from the British Library, London, UK.

Library of Congress Cataloging-in-Publication Data

Hooft, Katrien van t, 1956-
  Sustainable livestock management for poverty alleviation and food security / Katrien E. van't Hooft, Terry S. Wollen, Dilip P. Bhandari.
    p. cm.
  Includes bibliographical references and index.
  ISBN 978-1-84593-827-7 (alk. paper)
1. Livestock systems--Developing countries. 2. Food security--Developing countries. I. Wollen, Terry S. II. Bhandari, Dilip P. III. Title.

SF55.D44H66 2012
338.1'62--dc23

                  2011029852

ISBN-13: 978 1 84593 827 7

Commissioning editor: Sarah Hulbert
Editorial assistant: Alexandra Lainsbury
Production editor: Fiona Chippendale

Typeset by SPi, Pondicherry.
Printed and bound in the UK by the MPG Books Group Ltd.

Contents

Acknowledgements

The initial idea for this book was generated during the June 2005 International Workshop on Endogenous Livestock Development, held in Yaoundé, Cameroon. This workshop, hosted by Heifer International Cameroon, was an initiative of ETC-Compas, Heifer Netherlands and Agromisa – with the aim to understand better the practical implications of the Endogenous Livestock Development (ELD) concept.

This workshop was attended by 19 farmers from the north-west region of Cameroon – Fulani pastoralists, ethno-veterinary healers, dairy farmers and goat farmers – as well as staff members of Heifer Cameroon and government. International participation from India, Ghana, USA and the Netherlands contributed to the inter-cultural exchange and facilitation. Nine months later, two participants of the workshop made a tour to all involved participants and were surprised to find the positive effect of the ELD concept in their daily lives. This was documented in a book with a DVD in four languages, entitled *Endogenous Livestock Development in Cameroon – Exploring the Potential of Local Initiatives for Livestock Development*, published in 2008 by the Agromisa Foundation.

In 2008, the three authors of this book came together during a Central and Eastern Europe Area Program staff meeting of Heifer International in Vilnius, Lithuania. At this meeting, Drs Bhandari and Wollen presented the 'Heifer's Animal Well-Being Recommendations and Indicators' as part of the organization's Cornerstone on Improved Animal Management. This led to encouragement to put the guidelines into clear practice. This was the moment when the experiences of ETC-Compas, the ELD concept developed in Cameroon, and the Heifer International Recommendations and Indicators came together.

From there, and strengthened by the commitment from CAB International, the authors van't Hooft, Bhandari and Wollen started working on pulling these experiences together into this practical guide. This would not have been possible without the effective input from the participants of the various workshops, as well as others engaged in supporting ELD. Special thanks go to the authors of the case studies that are presented in Chapter 12.

Moreover, the manuscript would not have come together without the generous allowance of nights and weekends from our families, as well as the keen eye and diligent editing of Gayla Sybert from Heifer's administrative staff. The artwork for the two circle drawings are from the hand of Pooi Yin Chong, Senior Graphic Designer with Heifer. Finally, we are especially grateful to all farming families whose realities and experiences have one way or other contributed to the content of this book.

Introduction

This practical learning guide is about supporting livestock within the smallholder mixed farming systems around the world. It provides practical ways to understand and improve smallholder animal husbandry under a variety of agricultural and ecological conditions – based on the Endogenous Livestock Development (ELD) approach. *Agriculture* and *culture* are two systems that must work together for the benefit of men and women in rural and urban communities.

The title of this book could be *livestock in a fast changing world*. An estimated 70% of the world's rural as well as urban poor rely on livestock for their livelihoods. Two-thirds of all livestock are found in developing countries, while even developed countries find rural and urban poor livestock keepers with similar needs. Most smallholder farmers under these circumstances practise multi-purpose, low-input methods of livestock keeping.

All livestock and crop farmers today – especially those living in marginalized areas – are faced with rapidly changing socio-economic and ecological conditions. These include unexpected and recurring droughts, floods, extremes of temperature and other effects of climate change. Social changes overlie these environmental impacts because of out-migration of youth, larger production units owned by larger corporate units displacing the small farmer, land grabbing; all of this while there is a trend for improved markets for livestock products and by-products because of growing urban populations.

The writers of this training manual acknowledge that improving animal health and husbandry by rural and urban smallholder livestock keepers is not easily accomplished. All too often the multiple needs of the farmers are often not understood by donors and support providers, traditional practices are too often ignored and the right approaches to training are not taken. Farmers often return from training with new knowledge that does not fit their reality, or do not have the means to put the practices in place. It is commonly found that short courses and demonstration farms are not the total answer and often do not result in lasting change in animal health and production.

Thus, practitioners who provide training need to take a pause in order to see animal rearing through the eyes of the men and women farmers. Training in basic animal husbandry skills must start with that perspective and take into consideration the type of inputs as well as the (traditional) practices and innovations of those who rear the animals. Improved sustainable livestock management is much more than simply crossbreeding animals, providing feed and building shelters for the sake of increasing productivity.

Sustainable Livestock Management for Poverty Alleviation and Food Security is primarily written for livestock development practitioners working with limited-resource male and female livestock keepers: livestock development practitioners, animal health and husbandry trainers, extension workers, students in international agriculture and veterinary programmes, community animal health workers and veterinarians. It can be used in rural and urban small group settings and in the formal classroom.

The reader will recognize the two most common types of livestock keeping amongst smallholder families: (i) low-input and diversified livestock keeping; and (ii) more specialized smallholder livestock keeping. The understanding of the role, potential and limitations of each of these livestock keeping systems leads to different but complementary recommendations for supporting them.

The first part of the book (Chapters 1–5) provides a general overview of livestock development and smallholder livestock keeping of the rural (and urban) poor. In Chapter 1, the major trends in the livestock sector as a whole are presented. In Chapter 2, the various livestock development approaches are highlighted, explaining the most common causes of failure in livestock development projects and the potential of three relatively untapped sources of knowledge. In Chapter 3, the ELD approach is presented, highlighting the methodologies, organizations and networks that operate according to these principles. In Chapter 4, a differentiation between the four livestock production systems is highlighted, with smallholder farming and pastoralism the main livestock keeping strategies relevant to the urban and rural poor. In Chapter 5, the key elements of the smallholder livestock keeping systems, and the relevance to poverty alleviation and food security are presented.

In the second part of the book (Chapters 6–10) the two major types of smallholder livestock keeping systems are presented. The characteristics of smallholder low-input and diversified livestock keeping is described in Chapter 6. The characteristics of smallholder with more specialized livestock keeping in Chapter 7. Chapter 8 highlights the challenges and experiences in changing from smallholder low-input and diversified livestock keeping to more specialized livestock keeping. Chapters 9 and 10 present the complementary recommendations for supporting smallholder low-input and diversified (g) and more specialized (lo) livestock keeping.

Chapter 11 of the book is dedicated to experiences of finding pathways to markets, which can enhance smallholder livestock keeping systems. Chapter 12 presents a number of field-based cases of challenges and effective support to livestock keepers. Finally, in the Appendix an over-view of the recommendations in Chapters 9, 10 and 11 is presented.

We hope this practical learning manual will further enhance the effective support to smallholder livestock keepers. With adequate attention to the complex roles of animals within the smallholder systems, they can become an even more vital part of the family farm and provide positive benefits to farm resources. In turn, sustainable livestock production systems can protect the local environment while providing smallholder families with food, income and general well-being.

Katrien van't Hooft
Terry Wollen
Dilip Bhandari

1

Trends in the Livestock Sector

Learning Objectives: Understanding
- The critical issues confronting the livestock sector
- The main trends in the livestock sector and their background

The wise management of the world's agricultural biodiversity is becoming an ever greater challenge for the international community. The livestock sector in particular is undergoing dramatic changes as large-scale production expands in response to the surging demand for meat, milk and eggs.

Jacques Diouf, FAO Director General
(FAO, 2007)

One of the greatest challenges for humanity in the next decades is to maintain natural reserves in a world with a constantly increasing number of people. From the estimated 6 billion inhabitants in 1995, it is believed that there could be an increase of up to 8 billion around 2020, of which 40% will live in urban zones with a tendency of increased consumption of products of animal origin, especially meat. Global meat production tripled from 47 to 139 million tonnes per year between 1980 and 2002. As livestock production shifts into more high-input systems, it will place more pressure on arable land for the production of feed (FAO, 2008).

Today, most of the 640 million small-holders and 190 million pastoralists that are raising livestock are faced with rapidly changing socio-economic and ecological conditions. These include increased drought, floods and other effects of climate change, or social change because of out-migration of youth, as well as increased marketing opportunities for livestock products because of a growing urban population.

Three Critical Issues

Three critical issues are confronting the sector (FAO, 2009b):

Increasing pressure on ecosystems and natural resources – land, water and biodiversity

The livestock sector is only one of many sectors and human activities are contributing to the pressure. In some cases, its impact on ecosystems is out of proportion with the economic significance of the sector. At the same time, the sector is increasingly facing natural-resource constraints and growing competition with other sectors for a number of resources. Awareness is also increasing of the interactions between livestock and climate change, with the livestock sector both contributing to it and suffering from its impacts. Conversely, it is also being recognized that the sector can play a key role in mitigating climate change.

The globalization of food systems

An increasing flow of technology, capital, people and goods, including live animals and products of animal origin are moving around the world. Increased trade flows, along with the growing concentration of animals, often in proximity to large human populations, have contributed to increased risks of spreading of animal diseases and to a rise in animal-related human health risks globally. At the same time, inadequate access to veterinary services jeopardizes the livelihoods and development prospects of many poor livestock holders throughout the developing world.

Social implications of the structural changes in the sector

There are presently nearly one billion people who are undernourished. In facing this reality, more attention is being turned to smallholder farming. The social implications of the changes in food systems and the livestock sector need to be clarified. How can the livestock sector contribute more effectively to alleviating poverty and ensuring food security for all? Has the rapid development of the sector in many countries benefited smallholders, or are they being increasingly marginalized? If so, is this inevitable, or can the poor be brought into the process of livestock development?

General Trends

In addition to these critical issues, we can see the following trends in the livestock sector (FAO, 2009b):

- Large-scale production is rapidly expanding in response to surging demand for meat, milk, eggs and other products of animal origin. This is also known as 'the Livestock Revolution' and is based on rapid urbanization and increased purchasing power in many societies. The livestock sector is rapidly moving towards intensive and specialized systems, in which the production environment is controlled and aimed at maximum productivity per animal. Intensification is accompanied by the scaling up of production. This is combined with the globalized trade of animals and livestock products. These intensive and industrialized production systems contribute to meeting most of the growing demand for livestock-derived food. However, this focus on industrialization and globalization of livestock marketing also tend to influence the opportunities of poorer livestock keepers, often in a negative way.
- Increasing international trade as well as the rise of large retailers and integrated food chains are other important drivers of change in the livestock sector. This is a fairly recent phenomenon in developing countries. These include international market changes for animals and animal products; vertically integrated market changes, often created by foreign direct investments; influences of the local markets by globalization; and increasing local markets.
- Changing natural environments, such as recent changes in climate, have already affected biodiversity and ecosystems, especially in dryland environments. It is predicted that this degradation of ecosystems could be significantly worse during the first half of this century and be a barrier to achieving the Millennium Development Goals. The livestock sector will be affected, especially the pastoralist systems. The negative effects of climate change on extensive systems in the drylands are therefore predicted to be substantial.
- Technological advances are another driver of change. Advances in transport, cooling systems and communication

have promoted the expansion of global markets and facilitated the spread of production systems in which livestock are often raised some distance from their source of feed. Other technological advances include biotechnologies (artificial insemination and embryo transfer) and gene-based selection.

- Public policies as well as formal education and research are forces that add to the drivers presented above.
- The emergence of new and virulent animal diseases and their control measures has largely influenced the opportunities of small-scale livestock keepers in many countries. Examples include foot-and-mouth disease, hog cholera and more recently highly pathogenic avian influenza (HPAI, bird flu). Recent years have seen a number of serious epidemics, which have led to the loss of production and death of millions of animals. For example, the HPAI outbreak in 2003/2004 resulted in the slaughter of 30 million chickens in Thailand, amounting to approximately 29% of the country's native chicken population. Efforts to control the disease include the establishment of 'poultry-free zones' around large-scale production units, which inhibit the opportunities for local small-scale chicken production (see also Geerlings, 2007).
- Disasters and emergencies, such as drought, famine, floods, hurricanes, earthquakes, war and civil unrest, have devastating impacts on lives and livelihoods. The frequency of many types of disaster is increasing. The impact of such events on the livestock sector is not well documented. Food aid and restocking activities are not always appropriate to the local situation. Experience has shown that post-disaster restocking activities need to be carefully considered if they are to be well integrated and not to have an adverse effect on animal genetic diversity and the needs of the intended beneficiaries.
- Local animal breeds are threatened with extinction. A heavy reliance on just a few breeds of farm animal species, such as high-milk-yield Holstein-Friesian cows, egg-laying White Leghorn chickens and fast-growing Large White pigs, is happening while there is the loss of an average of one livestock breed every month. This is also called the 'Livestock Meltdown'. There is, therefore, an international call for *in situ* (farm raising) and *ex situ* (cryo-preservation) conservation activities of local, heritage and rare breeds.
- Average human food intake increases but hunger remains. According to FAO studies, under 'business as usual' the undernourishment in people's diet will decline from 20% in 1992 to 11% in 2015, but reductions in absolute numbers of undernourished people will be modest; from 776 million in 1992 to 610 million in 2015; far from meeting the World Food Summit target.
- The privatization of veterinary services in the developing world is taking place, whereas veterinary care, especially to the poorest livestock keepers in marginalized regions, is lacking. In the 1990s, when privatization was the buzzword, many countries commercialized the provision of livestock services, such as vaccinations, consultancy and training. This has had a major impact, especially on the poorer sections of society, and has given rise to many debates.
- Some argue that private markets serve people's individual needs best. For livestock services, this means that private providers are most efficient at delivering services such as artificial insemination and clinical veterinary care ('private good' services). Others argue that pushing veterinarians into privatization leads to less accountability and not more, because they are forced to practise 'health for profit' and not 'health for all'. Democratizing the services would involve decentralized governance, appropriate extension work, prevention, accountability and transparency to farming communities. This demands greater public investment and not less, to enable a more effective and farmer-owned 'free' service (Ahuja and Ramdas, 2010).
- The commercialization and over-exploitation of medicinal plants and

the growing interest in alternative treatments has resulted in the hunt for potentially useful indigenous knowledge. Firms and research groups seek to patent ingredients or preparations, and remedies are commercialized by outsiders, often bypassing the communities that developed them. These developments have resulted in the overuse of selected medicinal plants, while local communities have rarely benefited from the use of their knowledge and plants by outsiders.

- There is increasing concern and criticism on uncontrolled market forces, both in developed and developing countries and the effect of the above-described trends in relation to the environment, poverty alleviation and MDGs, animal genetic resources, and animal wellbeing, to name but a few. This concern comes from civil society organizations as well as UN organizations and governments. The FAO of the UN, for example, is designing global guidelines to secure valuable animal genetic resources. Alternatives are posed in many ways, e.g. in the form of the slow food movement, consumer–producer alliances, (re-)diversification of agriculture and endogenous development. In higher policy levels, and in spite of the concern expressed, these alternatives are not yet taken into account in a serious way.

Other Trends

Land grabbing

Since the beginning of 2009, large-scale acquisitions of farmland in Africa, Latin America, Central Asia and South-east Asia have made headlines in a flurry of media reports across the world. Lands that only a short time ago seemed of little outside interest are now being sought by international investors to the tune of hundreds of thousands of hectares. This is a hot issue because land is so central to identity, livelihoods and food security. Several factors seem to underpin these land acquisitions. These include food security concerns, particularly in investor countries, which are a key driver of government-backed investment (Cotula et al., 2009). Government-backed deals can also be driven by investment opportunities rather than food security concerns.

In addition, global demand for biofuels and other non-food agricultural commodities, expectations of rising rates of return in agriculture and land values, and policy measures in home and host countries are key factors driving new patterns of land investment (Mathias, 2007).

For people in recipient countries, this new context creates risks and opportunities. Increased investment may bring macro-level benefits (such as GDP growth and improved government revenues), and may create opportunities for economic development and livelihood improvement in rural areas. However, as governments or markets make land available to prospecting investors, large-scale land acquisitions may result in local people losing access to the resources on which they depend for their food security – particularly as some key recipient countries are themselves faced with food security challenges.

While there is a perception that land is abundant in certain countries, these claims need to be treated with caution. In many cases, land is already being used or claimed – yet existing land uses and claims go unrecognized because land users are marginalized from formal land rights and access to the law and institutions. Even in countries where some land is available, large-scale land allocations may still result in displacement as demand focuses on higher value lands (e.g. those with greater irrigation potential or proximity to markets).

Expansion of rights-based approaches

Livestock Keepers' Rights and Bio-cultural Community Protocols (BCPs) have come up over the past few years to provide livestock-keeping communities the opportunity to document and showcase their role in the management of animal genetic resources and agro-ecosystems. Indigenous and local livestock keepers have been recognized as 'Guardians of biodiversity' (FAO, 2009a). Community Protocols (CPs) are increasingly recognized by biodiversity conventions, and are now part of the Nagoya Protocol on Access and Benefit Sharing that was adopted at the COP 10 of the UN Convention on Biological Diversity in Japan, October 2010. This means that within 2–5 years the national governments will start using CPs in their strategies to promote biodiversity. BCPs are promoted to include the cultural elements in this effort (LPP and LIFE Network, 2010; Compas, 2010).

Increased problems with multi-resistant microbe strains

A series of complicated problems with the resistance of bacterial pathogens to commonly used antibiotics has evolved, especially within the intensive livestock production systems. A part of this is blamed on the widespread use of antibiotics for disease prevention and growth promotion. In some countries, it is alleged that intensive pig production has led to the development of multi-resistant *Staphyloccus aureus* (better known as MRSA) in the human population and has caused major problems and costs within hospitals. The use of antibiotics for growth promotion is now prohibited within the EU. More recently, a multi-resistant component that can move from one microbe to the other – also called extended-spectrum beta-lactamase-producing microbes or ESBL – has been identified, which is linked to excessive antibiotic use in intensive chicken production. This threat is now also growing on a worldwide scale (Kumarasamy *et al.*, 2010). Indiscriminate use of antibiotics in smallholder systems contributes to further this problem.

Renewed interest in traditional practices – ethno-veterinary medicine

Around the globe, traditional animal health care approaches are having a comeback, triggered by an increasing recognition of their value and the need to reduce the heavy use of commercial products in veterinary medicine and agriculture. In 1986, McCorkle coined the term 'ethno-veterinary medicine' for such approaches and – together with other scientists and development practitioners – advocated for their enhanced use in development. Since then, the number of studies, projects, documents and theses on ethno-veterinary medicine has been steadily increasing, both in developing and developed countries (FRLHT and Tanuvas University, 2010). A comprehensive book on *Ethno-veterinary Botanical Medicine* was published in April 2010 (Katerere and Luseba, 2010).

However, veterinary faculties have been slow in picking up on this trend, offering information on medicinal plants at best. Ethno-veterinary medicine has much more to offer. A systematic integration of ethno-veterinary practices and information into veterinary curricula and education would not only widen the spectrum of prevention and treatment choices, it could also deepen the understanding of and respect for health care approaches of communities and make veterinary services in marginal areas more appropriate to the needs of livestock keepers. Plus, it would cater to the increasing demand among clients for alternative health care approaches (Mathias, 2010).

One Health – One Medicine Approach

It is becoming much more evident that human health, animal health, public health and environmental health are all connected (ILRI, 2008). This recognition has led to the emergence of an initiative that is variously called as One Health or One Medicine, and is bringing practitioners of each field together. One Health may seem far removed from the low-input livestock systems; however, it has immense implications in villages with limited resources, especially related to the control of zoonotic diseases. When veterinary and human health services join forces, for example, vaccinations

against rabies can be more efficient, cover-
ing dogs and cattle in one community at the
same time (Be.Troplive, 2010).

Zoonotic diseases might take epidemic
(e.g. rabies, Rift valley fever) or endemic
forms (e.g. brucellosis, echinococcosis).
Whereas emerging and epidemic zoonoses
usually attract much interest, endemic
zoonotic diseases rarely give rise to
collaboration between the medical and veterinary professions, especially in developing countries. In fact, these are the neglected
zoonotic diseases, officially recognized as
such by the World Health Organization
(WHO, 2005). There is a clear need for
more intense collaboration between human
and animal health professionals in the
control of these diseases.

References and Further Reading

Ahuja, V. and Ramdas, S.R. (2010) Livestock services to family farmers: free or fee? *Farming Matters*, March.
Be.Troplive (2010) *Invitation Symposium 'Where medics and vets join forces'*. Institute of Tropical Medicine
    (ITM), Antwerp, 5 November.
Compas (2010) Bio-cultural Community Protocols enforce Biodiversiy Rights – a selection of cases and expe-
    riences. *Endogenous Development Magazine* no. 6, July.
Cotula, L., Vermeulen, S., Leonard, R. and Keeley, J. (2009) *Land Grab or Development Opportunity?
    Agricultural Investment and International Land Deals in Africa*. IIED/FAO/IFAD, London/Rome.
FAO (2007) *The State of the World's Animal Genetic Resources for Food and Agriculture*. FAO, Rome.
FAO (2008) Livestock in a changing landscape. UNESCO/Scope/UNEP policy brief no. 6, April. FAO, Rome.
FAO (2009a) Livestock keepers – guardians of biodiversity. Animal Production and Health Paper no. 176.
    FAO, Rome.
FAO (2009b) *The State of Food and Agriculture – Livestock in the Balance*. FAO, Rome.
FRLHT and Tanuvas University (2010) *Proceedings of International Conference on Ethnoveterinary Practices.
    Mainstreaming Traditional Wisdom on Livestock Keeping and Herbal Medicine for Sustainable Rural
    Livelihood across Continents*. Thanjavur, Tamil Nadu, 4–6 January.
Geerlings, E. (2007) Highly pathogenic avian influenza: a rapid assessment of the socio-economic impact on
    vulnerable households in Egypt. FAO/WFP Joint Project Report.
ILRI (2008) Nine myths about livestock in developing countries. Adapted from: Randolph, T.F., Schelling, E.,
    Grace D., Nicholson C.F., Leroy J.L., Cole D.C., Demment M.W., Omore A., Zinsstag J. and Ruel M.
    (2007). Invited review: Role of livestock in human nutrition and health for poverty reduction in develop-
    ing countries. *Journal of Animal Science* 85, 2788–2800.
Katerere, D.R. and Luseba, D. (2010) *Ethno-veterinary Botanical Medicine – Herbal Medicines for Animal
    Health*. CRC Press, Boca Raton, Florida.
Kumarasamy, K.K., Toleman, M.A., Walsh, T.R., Bagaria, J., Butt, F., Balakrishnan, R., et al. (2010) Emergence of
    a new antibiotic resistance mechanism in India, Pakistan and the UK: a molecular, biological, and epide-
    miological study. *The Lancet Infectious Diseases* 10, 597–602.
LPP and LIFE network (2010) *Biocultural Community Protocols for Livestock Keepers*. Lokhit Pashu-Palak
    Sanstahn (LPPS), Sadri, Rajasthan.
Mathias, E. (2007) Is biofuel the answer? Experiences of pastoralist and smallholder livestock keepers.
    Powerpoint presentation at VSF-Europe Meeting Portugal. LPP and ELD network.
Mathias, E. (2010) Mainstreaming ethnoveterinary medicine in veterinary education and research. In:
    Proceedings of the International Conference on Ethnoveterinary Practices, Thanjavur, Tamil Nadu, 4–6
    January.
WHO (2005) The control of neglected zoonotic diseases – A route to poverty alleviation. Report of a Joint
    WHO/DFID-AHP Meeting, Geneva, September.

2

Livestock Development Approaches

Learning Objectives: Understanding
- The differences in livestock development focus
- The animal productivity paradigm
- The farm efficiency and integrated farming paradigm
- Where did it go wrong? Reasons for project failures in livestock development
- How climate change and food security call for a new approach
- Three levels of relatively untapped potential in livestock development

On Whom to Focus in (Livestock) Development?

In this book, we especially focus on smallholder mixed-farming system. Within this system, we will distinguish between the low-input and diversified smallholder livestock keeping (Chapters 7 and 9) as well as on more specialized smallholder livestock-keeping activities (Chapters 8 and 10).

In (livestock) development, there is a wide array of views on the best way to achieve a positive outcome, in terms of food security, income generation, employment and (environmental) sustainability. Some argue that funds and efforts are best spent on specialized (middle-level) commercial farmers that have the capacity to hire other people as labourers; others argue that poor smallholder farmers and pastoralists need to be supported directly. As an example, in Table 2.1, the various arguments on this topic are presented,

which are from a discussion between four development professionals during the 2006 German Tropentag in Hohenheim, Germany.

The Animal Productivity Paradigm

Most livestock policies, education and research in developed and developing countries are based on the animal productivity paradigm, common in the high-input commercial livestock production systems. The main aim within the productivity paradigm is to produce the highest amount of produce (milk, meat, eggs, honey, etc.) per animal per day, at the lowest monetary costs.

In itself, specialized livestock production based on the animal productivity paradigm is good news, especially in view of the need to provide inexpensive food for urban populations. However, politicians, donors and investors often prefer big, prestigious

Table 2.1. Arguments for focusing on different farmer groups.

Topic	Arguments to focus on middle-level farmers (more specialized)	Arguments to focus on smallholder farmers (low-input and diversified)
Which farmers to support?	We have to work with middle-level farmers, e.g. use a threshold over 10 ha, or farmers that can hire labour. Hunger has to do with purchasing power. People need money and employment. Therefore, development for marginal farmers is only possible through enhanced off-farm income generation.	We need to focus on the potentials of small producers, e.g. by using micro-credit.
Rural development or poverty reduction?	The objective is poverty reduction in rural areas, not rural development. The exclusive focus on agriculture can lead to a poverty trap. Poverty reduction requires e.g. schooling for girls, rural roads, etc. It is rare to hear about rural development in the international discussions.	Policy and research needs to be adapted to the marginal rural areas, where rural development is necessary – research needs to focus on low-potential areas. Governments have diminished support for marginal rural areas; there has been discrimination of the poorest farmers.
Local markets	Local markets cannot be depended upon; they fluctuate too much to stand as the basis of income generation.	We need to guarantee and defend local and regional markets, improve their products, enhance food storage and processing.
Global food chains and smallholder farmers	Global food chains are a major threat to smallholder farmers. Marginalization is increased by globalization. Poor farmers are not likely to participate in the globalized food chain. Small commercial farmers might be more successful.	We need to consider whether we want to invest in global food chains or regional ones. Modernization does not work for the poorest. Smallholder farmers have problems with economy of scale. International requirements are a disadvantage to smallholder farmers.
Work with marginal farmers	For marginal farmers, we have to stabilize their subsistence production. The only opportunities for smallholder farmers are their relative advantage in labour-intensive production activities. There is a need for technology transfer, e.g. related to land tenure, property rights and increased production of smallholder farm enterprises.	We have to decrease vulnerability, e.g. through direct marketing, payment for environmental services, tourism etc. Identify innovations at the local level. People often do not pick up an innovation in the way it was intended, but these are often used in other ways (e.g. toys made from cola tins). We also need to guarantee access to land and the tenure of property rights.
Policy level	At policy level, we have to create a 'level playing field' that makes it possible for smallholder commercial farmers to participate. This requires supply chains and support to farmers' organizations, etc. There is a need for a public–private interface, but there are difficulties in the interaction.	Smallholder farmers need a functioning system to participate in decision making.

projects rather than building on local initiatives. Decisions about livestock development are often made in capital cities, not in villages or pastoral encampments. Not everyone benefits from these intensive, specialized livestock production systems.

The social and environmental costs are most often not included in the calculations and are often difficult to agree upon. Development based on the productivity paradigm includes the premise that high-input and specialized animal production systems – especially intensive poultry, pig and dairy farms – will provide inexpensive, high-quality and abundant food for the population, as well as a livelihood for the farmers involved. These policies are primarily aiming at supporting specialized, commercial and high-input farming operations.

The high-input livestock production systems are growing at an unprecedented rate in developing countries as well as in other parts of the world. These are sometimes supported by large international companies with major economic and political power, such as the fertilizer and chemical industries, genetic industries such as Monsanto, as well as the soy-lobby and the lobby of animal industries, to name but a few. This process, also called the Livestock Revolution, has numerous positive as well as negative impacts on smallholder agriculture (Mathias and Mundy, 2008).

The Optimizing Farm Efficiency and Integrated Farming Paradigm

The livelihoods of millions of small-scale families in marginal areas of the developing world depend on livestock, in mixed farming systems, (agro-) pastoralism systems or agro-forestry systems that include several livestock species.

FAO statistics (FAO, 2009) indicate that 69% of global agricultural land and 26% of total land is covered by agricultural pastures, rangelands and grasslands. Mixed farming and agro-pastoralist systems occupy a potential of some 2.5 billion ha of land, of which 1.1 billion ha are arable rain-fed cropland, 0.2 billion are irrigated cropland and 1.2 billion ha are grassland.

As most of the rangelands/grasslands/ pastures are too dry, too wet, too cold or too high to be cropped, these areas can be used agriculturally only through agro-pastoral livestock production, which supports several hundred million people.

In developing countries, the majority of rural households are smallholder crop-livestock farmers. Livestock are an integrated part of their agricultural systems that stand as the basis of the farm and family life in many parts of the world. In most rural communities in Asia, Latin America and Africa, livestock are essential for draft power, organic fertilizer and transportation. They are a source of nutritional protein, wool, leather and fibres. Raising animals is also a form of security against climatic and economic risks, and a means of accumulating and maintaining financial reserves.

In this way, animals optimize the efficiency of the farm as a whole, in terms of soil fertility, energy efficiency and minerals such as nitrogen and phosphorus. This is increasingly perceived as an option for food security and poverty alleviation, as well as climate change adaptation and mitigation.

This *integrated farming paradigm* is also being recognized within high-input dairy in developed countries, such as in The Netherlands. It is based on optimizing the efficiency of the dairy farm as a whole, and on increased soil fertility and natural resistance of soils, animals and plants. In this way, fewer inputs in terms of commercial fertilizer and pesticides/herbicides are needed, as well as reduced input of concentrate feeds. This increases farm efficiency, both in terms of energy efficiency and of soil minerals such as nitrogen and phosphorus, and stimulates farmer income. This is increasingly perceived as an option for food security, poverty alleviation, and climate change adaptation and mitigation (see Box 2.10 at the end of this chapter).

Where Did it Go Wrong?

Male and female farmers on marginalized lands with limited capital, formal education and opportunity continue to depend on low-input agricultural practices. In supporting livestock development in these areas, however, these diverse and low-input farming strategies are often perceived as a problem rather than a potential. This lack of understanding of the socio-economic and cultural reality of the families, the role of the animals within their integrated farming system, combined with top-down methodologies, has resulted in a disappointing impact of livestock development efforts in terms of poverty alleviation (Box 2.1).

Some of the causes of lack of impact found with greater frequency are:

1. Changing production systems. Projects tend to change the production system for another, generally from a low-input and diversified system to a more specialized system, without taking into account the social implications and risks of this process for the families involved. In addition, they want to implement activities and technologies based on the productivity paradigm without considering aspects of farmer knowledge and (ethno-veterinary) practices of the people involved.

2. Lack of social and cultural sensitivity. In many projects, a thorough analysis of the reality of the families involved is missing, especially in relation to their ways of perceiving the world (worldviews), the logic of their local (integrated) production systems, the function of animals within this reality and how the families perceive the project. This lack of social and cultural sensibility may have to do with the changing and abstract development approaches shown in Fig. 2.1.

3. Over-ambitious project objectives. Often the objectives of the projects are too ambitious, including too much territory or too many communities. In these cases, there is an imbalance between activities established by the project and what can concretely be achieved. Thus, in practice, the achievements in each one of the communities can be limited, often by logistical reasons such as transport, but also by factors such as unfavourable climate, lack of institutional support and personal problems of the extension workers involved.

4. Not including women. Many projects are mainly directed by men, whereas those principally responsible in animal husbandry, especially in low-input diversified husbandry, are women (Kristjanson *et al.*, 2010). The difficulties in establishing relationships between the extension workers and the women are great. When the extension workers speak the native language of the population, their relationship with the women is generally considerably easier.

5. Top-down methodologies. Personal ideas and the education of the extension workers sometimes clash with the reality of the farm families, affecting the exchange of ideas and dialogue necessary to establish a true participatory process. In the planning phase, even with 'participatory activities', the basic ideas generally come from the extension workers of the implementing organization rather than from the farming community members themselves.

Box 2.1. Review of livestock development efforts.

A somewhat dated but comprehensive review of 800 livestock projects (Livestock in Development, 1999) concluded that many projects have had disappointing results in terms of poverty alleviation. The diffusion and adoption of the livestock technologies promoted by governments and non-governmental organizations (NGOs) so far has been very limited and very often they have not helped the poor. The report attributed this to the lack of focus on poverty. In most cases, the technologies offered were not appropriate and were imposed on people using a top-down approach, without mobilizing their own strengths and resources, and without regard for existing traditional knowledge and institutions. The approach fails to recognize the multi-faceted role of livestock for human society, and ignores other dimensions.

Fig. 2.1. Different realities: top left, reality perceived by the farmer; top right, what the interdisciplinary team perceives; centre left, the vision of the industrial dairy specialist; centre right, the agronomist's vision; bottom left, the sociologist's vision; bottom right, the ecologist's vision (van't Hooft, 2004).

6. Lack of practical experience. Several organizations, especially NGOs, have a gender department. In the case of income-generation projects for women, livestock activities with pigs, chickens or guinea pigs are frequently initiated. In addition, they are often communal projects. These projects have confronted many problems, such as the lack of technical and economic experience, unexpected changes in the male–female relationships because of the project, and the difficulties related to a communal farm.

7. Power differences in managing funds. The projects have to render accounts of their effect to donors, state organizations and NGOs rather than to the population's own organizations. The relationship between the project, the extension workers and the families involved implies differences in power, especially relative to the management of funds. As a result, there is a lack of real coordination between the extension workers and the families that finally resign themselves to 'whatever comes of it' instead of taking charge of the project.

8. Communal livestock projects. There are many ideas and ideologies related to the support of the communal organization via community livestock projects. The success or failure often depends on previous experiences and organization structures. Community organization is more frequent within pastoralist societies, for example the joint shepherding of sheep or llamas among groups of families in the Andean region. Amongst smallholder farmers, however, the husbandry of cows and other species such as poultry, pigs or guinea pigs is generally done near the house. For this reason, many communal projects with these species in smallholder farmer communities have failed (Box 2.2).

9. Introduction of exotic breeds. The introduction of exotic breeds is widespread in

Box 2.2. Failure of communal chicken project in Nicaragua (van't Hooft, 2004).

In an indigenous town on the Atlantic coast of Nicaragua, some 10 h by boat from a town with a commercial market, an extension worker from a local NGO designed a communal poultry project with exotic laying hens, balanced feed and the other elements of specialized chicken production. The women of the community constructed a communal chicken farm and initiated the husbandry project. In a short time, the animals began to die from cannibalism because – when the expensive shipments of balanced feed did not arrive on time – the women began to substitute with local feedstuffs. This caused under-nutrition and stress in the high-productivity hens. The women gradually moved away from the project. When the women who had continued arrived at the conclusion that the best solution was to divide the hens and to include them into their home husbandry activities, the extension worker of the NGO 'prohibited' such action. He promised to ship in another load of balanced feedstuffs. The communal chicken farm failed several months later.

livestock development projects in developing countries. There are numerous reasons for the frequent failures – the introduction of animals of exotic breeds, such as Holstein dairy cows and Corriedale sheep, for example. Sometimes these new breeds are introduced without the existence of adequate husbandry conditions (Box 2.3) or motivation on the part of the families to change the system of husbandry to a more intensive system. Numerous examples have shown that animals of exotic and specialized breeds, upon being introduced to their new environment, are confronted with a series of limitations that they cannot cope with because of their sensitivity to lack of food, diseases and parasites.

10. Not including animal species relevant to the poor. The livestock-keeping systems of poorer households tend to include a wide number of livestock species with the aim of minimizing and spreading risks. For example, most rural families keep poultry, but that is not limited to chickens only, as chickens are highly susceptible to the common and fatal Newcastle disease. It is common to find poultry flocks that include ducks, turkey and guinea fowl. Other fowl species used are doves and quails. These animal species often provide culture-specific products, dishes and services. Most poultry projects, however, are limited to the most commonly used animal species in intensive animal production: chickens.

11. Micro-credit linked to improved animal health practices. Efforts to increase income of poor rural farmers through micro-credit loans are often directly or indirectly linked to improved animal health practices that require higher inputs. Many of these programmes introduce Green Revolution technologies, such as crossbreeding with highly productive breeds, use of commercial fertilizer, improved seeds and other commercial inputs. Painful experiences have shown that long-term effects can be quite the opposite, especially in environmentally risky areas. These measures have frequently led to serious environmental degradation, the genetic loss of resistant local breeds, and high vulnerability and financial drawbacks to the families involved.

12. Disease control depopulation schemes (Geerlings, 2007). Killing of flocks and other individual animals in the face of disease control programmes is another element that has resulted in negative experiences, especially of the most vulnerable livestock-keeping families. Examples are the eradication of poultry in the avian influenza outbreaks between 2006 and 2007, and hog cholera/African swine fever outbreaks. Broad-spectrum disease control practices need to include components that are designed for the smallholder as well as commercial producers.

> **Box 2.3.** Example of exotic breed that failed to adapt to its new environment (van't Hooft, 2004).
>
> In the humid tropics of Bolivia, the introduction of white pigs, such as the Yorkshire and Landrace breeds, has not worked. One of the adverse factors has been the existence of vampire bats which attack and suck blood. Because of this, the female pigs may lose their nipples before the first birth and consequently they lose their ability to raise their young. This problem affects the local breeds with red or black skin less than the exotic white-skinned breeds.

Climate Change and Food Security Call for a Renewed Approach

Rarely does one see images of farmers ploughing fields, planting seeds or feeding animals in relation to climate change. Indeed, until recently, agriculture – particularly in developing countries – has been largely absent from climate change and food security discussions. However, farming is intimately involved in climate change. Agricultural activities, including forest clearing, fertilizing soil and transporting produce, and indeed livestock farming, contribute significantly to global greenhouse gas emissions (FAO, 2006).

Meanwhile, farmers, particularly in developing countries, are also the major victims of climate change. They are threatened most by climatic changes such as shifting rainfall patterns and more extreme and unpredictable weather.

As Carlos Seré, Director General of the International Livestock Research Institute (ILRI) in Kenya, reported in his paper, 'No Simple Solutions to Livestock and Climate Change' (2009):

> Livestock emissions depend, however, on how animals are raised and fed. Grain-fed, factory-farmed cattle in industrialized countries emit much higher levels of greenhouse gases than the grass-fed, family-farmed cattle in developing countries. Most people who keep cattle in developing countries are either small farmers who feed their animals available vegetation with seasonal supplements of stalks and other harvested crop wastes, or herders who periodically move their stock in search of new sources of grass and water. Both these groups have very few

alternatives for making a living beyond crop and livestock farming and both leave a relatively small environmental footprint. For example, all of Africa's cattle and other ruminants contribute just three per cent of global livestock methane emissions.

Achim Steiner, the Executive Director of the United Nations Environment Programme, presented the role of sustainable agriculture for climate change mitigation during the EU Agriculture and Climate Change conference in Brussels:

> We should not only invest in high-technological solutions, but rather invest in smallholders farmers. We have to do this. Without attention to agriculture and food security there can never be an agreement with developing countries in Copenhagen.
>   (article Volkskrant, 27 June 2009)

Olivier de Schutter, the UN Special Rapporteur on the Right to Food, also presented this view during a debate in the series 'The Future of Agriculture and our Food' (Rode Hoed, Amsterdam, 10 November 2009):

> The UN now recognizes that it has been a mistake to exclusively support large agricultural enterprises. The Green Revolution model has produced more food and more hunger at the same time. Alternatives are silenced, not taken serious or widely under-estimated. In reality, agro-ecological farming is extremely productive per hectare. It is necessary to re-invest in smallholder agriculture.

The required changes are those proposed in agro-ecological farming and sustainable

---

Box 2.4. Livestock and climate change: are win–win effects possible? (van't Hooft, 2009).

Livestock are increasingly being cited as one of the major producers of greenhouse gases: some reports even indicate a contribution of 51% of the total of greenhouse gases produced. Replacing livestock products with meat and dairy analogues based on soy, rice or wheat is suggested as the most desirable way out. Unfortunately, reality is more complex than this. Livestock is not produced in one way, which can simply be replaced. Livestock emissions largely depend on how animals are raised and fed. Grain-fed, factory-farmed cattle emit much higher levels of greenhouse gases (and other environmental effects) than the grass-fed, family-farmed cattle, although their emissions per kg of milk produced is lower.

Fortunately, other international reports (IPCC, 2007; FAO, 2009) indicate another way out: increased sequestration of soil carbon through sustainable use of soils and other resources in agriculture. These reports estimate that 90% of the total mitigation could come from sink enhancement (soil carbon sequestration) and about 10% from emission reduction. Although not explicitly mentioned in the reports, this puts livestock in a different perspective.

Different livestock species are farmed as an integrated part of mixed farming systems, agro-forestry systems, pastoralist and agro-pastoralist systems throughout the world. There is an enormous scope for building on the experiences gained by supporting soil fertility within both smallholder systems as well as industrialized livestock-keeping systems. This can have a surprising win–win effect in terms of both food security and climate change.

---

land management: improved cropland management, water management, pasture and grazing management, restoration of degraded lands, and management of organic soils.

In this process, there is a need to look at various animal production systems, both high- and low-input systems (Chapter 4). There is an enormous scope in building on the experiences gained in supporting sustainable land management within each of these systems (Box 2.4). This can have a surprising win–win effect in terms of both food security and climate change.

More recent reports on the link between food security and climate change (FAO, 2010a) call for a Climate Smart Agriculture, with a similar ecosystem approach: 'The overall efficiency, resilience, adaptive capacity and mitigation potential of the production systems can be enhanced through improving its various components'.

A monitoring report of climate efficiency of mixed smallholder systems in Uganda commissioned by Send-a-Cow indicated that positive outcomes are possible (Alford and Penney, 2006).

These promising views are not always translated into livestock development practice, however. Recent documents from the livestock sector – especially the dairy sector

(FAO, 2010b) – indicate that (sustainable) intensification of livestock production is the only way to reduce the emission of greenhouse gases from the livestock sector. It remains to be seen what effective measures will be taken in practice to promote this sustainable intensification: based on an integrated 'smart-climate' approach or rather on the conventional 'increased animal-productivity approach'.

Three Relatively Untapped Resources

Most agricultural policies and curricula still tend to focus on the productivity paradigm within high-input (semi-)industrial livestock production systems, and fail to differentiate their strategies according to the animal production system. They tend to promote replication of standard knowledge rather than stimulating the creative thinking needed to confront the major social and environmental challenges ahead. The agro-pastoralist and mixed farming systems tend to receive little attention and support within education and government programmes. It leaves problems unanswered and it underutilizes their unique potential.

One can state that there are three relatively untapped resources that can stand at the basis of new initiatives in sustainable (livestock) development: (i) lessons from innovative farmers; (ii) lessons from successful field-level organizations; and (iii) lessons from intensive livestock keeping in developed countries.

Lessons from innovative farmers

Mostly poor farmers in agro-pastoralist and mixed farming systems are less capable of profiting from increased demand of animal products in global markets and they have less access to 'general solutions', such as privatized animal health systems. They do have specific local challenges, however, and they represent an opportunity to develop local solutions to local problems while making a living by producing food and adapting to climate change (Boxes 2.5 and 2.6). Their obstacles are many, such as degraded natural resources, lack of financial resources, access to new skills and knowledge, weak institutions, inadequate infrastructure and poor governance. With proper approaches, however, this is a relatively untapped potential in the process of adapting to the various ecosystems.

The large variation in these systems together with their unique combination of livestock, crops, soils and society can provide clues for sectors other than only livestock production. Moreover, the integrated nature of their soil–plant–animal–people systems can have low CO_2 footprints and positive effects on many Millennium Development Goals (MDGs), while also providing options to adapt to climate change.

Lessons from successful field-level organizations

There is a second untapped potential for livestock development available. This lies in successful service providers to these farmers, such as NGOs, farmer organizations, private enterprise and educational institutions – organizations that managed to combine formal education with successful field-based work to 'produce' professionals that are better equipped to work in sustainable development. Such service providers have developed methodologies to cope with the specific challenges in these systems. They are based on time-tested relations with their farmers, use of local skills and resources, and balanced input of external resources within local realities and culture (Boxes 2.7, 2.8 and 2.9). Much of this work is relatively unknown outside their direct development circle,

---

Box 2.5. Innovation from a South African farmer: offering leafy branches (Letty and Waters-Bayer, 2009).

In Msinga, KwaZulu-Natal (KZN), many households have indigenous goats, and women in the household are often involved in managing them. The goats go out to graze during the day and must be brought home in the evening to ensure that they are not stolen or taken by predators. Because the goats must cover long distances to be able to find enough feed, a great deal of time is often needed to find them and bring them home in the late afternoon.

One of the farmers encountered through the process of documenting local innovation was Mrs Maduba Mbila (Fig. 2.2). She had developed an effective means of ensuring that her goats returned home every evening without household members having to go and fetch them. She offers them various palatable leafed branches (e.g. *Schotia brachypetala*) and water when they return to the kraal. She discovered this mechanism by chance. She had kept several female goats at home while their kids were small and fed them with leafy branches lopped from indigenous trees and bushes growing naturally in the vicinity of the homestead. When the kids became older and she released the female goats for grazing, she found that they continued to return home in the evening and brought the rest of the flock with them.

This innovation has proved very useful for Mrs Mbila, as it has reduced the effort and time needed to ensure that the goats are kraaled at her home every night.

Continued

Box 2.5. Continued.

Fig. 2.2. Mrs Maduba Mbila in Msinga, KwaZulu-Natal Province uses palatable leaf branches to attract her goats home each evening. This reduces the time spent fetching them and losses from predators. Credit: Prolinnova South Africa.

Box 2.6. Another innovation: raised grass baskets (Letty and Waters-Bayer, 2009).

Mrs Mbuyisa keeps backyard chickens and has developed a system of raised grass baskets in which her hens lay eggs. This has proved effective as a way of making it easier to find and collect the eggs. The extension staff from the Mpumalanga Department of Agriculture are planning to assist her with growing supplementary feed for the chickens in an effort to prevent them from wandering out of her yard in search of food.

Mrs Dlamini is a member of a community that makes chicken nesting boxes out of the bases of sisal stems. The stem is stripped of leaves and hollowed out to create a nest. In addition, the women have found through informal experimentation that, if some burning grass is used to burn off the inside of the hollowed nest, a smoother surface is obtained, which creates a less favourable environment for external parasites. They also found that the nests insulate the chickens well and protect them from predators.

but this potential can be used more effectively, by strengthening mechanisms for exchange and joint advocacy.

Lessons from intensive livestock keeping in developed countries

Solutions found in intensive livestock production systems in the so-called developed countries, which have faced environmental challenges, can provide relevant options for farmers and organizations elsewhere. Examples are the dustbowl of the central US plains in the 1930s, Southern Australia in the 1980s, and the acid rain problems related with pollution caused by excessive use of inputs in industrial livestock systems in Western Europe. Experiences from these cases show that farmers with or without the official infrastructures can both

---

Box 2.7. Successful fleece improvement programme of local sheep (Perezgrovas, 2006).

In Highland Chiapas, the Tzotzil ethnic group of Mayan origin derives up to 36% of its income from sheep husbandry and from the weaving of typical woollen clothes; government efforts have attempted to absorb the local wool sheep with high-producing breeds such as the Merino, without success. A different approach was attempted to improve the quality of wool in the local Chiapas sheep by means of animal selection, and a research project was designed utilizing an open nucleus scheme.

Commercial or industrial traits of high-quality wool (white, short, fine) were exactly the opposite of those developed by the local weavers (coloured, coarse, double-coated, long). To account for the difference, groups of Tzotzil shepherdesses and weavers were invited in 1996 to collaborate as part of the sheep-improvement plan, directing research goals by means of their continuous assessment of fleece quality in the animals of the nucleus flock. This collaboration is put into practice by grading the quality of the fleece in all sheep under 24 months of age, prior to each 6-monthly shearing. The list of achievements in the first 10 years of this unique inter-ethnic collaboration includes a set of selection objectives for fleece quality, and a comprehensive understanding of the characteristics of wool in the local sheep and its relationship with the transformation of wool into clothes through an ancient textile process utilized over centuries by the Tzotzil women (Fig. 2.3).

As a result, current fleece variables within the improvement programme include: fleece quality, staple length, textile aptitude (proportion of coarse/fine fibres), greasy fleece weight and wool growth. Improved rams from the nucleus flock have been introduced within community flocks, and their offspring have inherited superior fleece-quality traits.

Fig. 2.3. Improvement of local breeds of sheep by the Institute of Indigenous Studies of the University of Chiapas, Southern Mexico, is done in close collaboration with the Tzotzil women that keep them. Selection is done on the basis of the criteria of the shepherdess. Credit: Raul Perezgrovas.

Box 2.8. Promoting local salt-licks (Karbo, 1999).

In northern Ghana, farmers traditionally use local salt-licks from a specific naturally occurring salty soil, called siella. Local farmers experimented by adding other nutrients to this soil, especially oyster shell and cassava flour. Scientists from the Animal Research Institute (ARI) and development workers from the Association of Church Development Projects (ACDEP) and MoFA jointly conducted on-farm trials of a mineral lick made from bone ash and salt. At community meetings, farmers evaluated the results as being useful for their sheep and goats, thus confirming the findings from the on-station trials.

This resulted in a salt-lick that their animals love and thrive well on. Farmers observed that using the lick at home made it easier to manage the animals, as they returned to the pens early to receive the lick. Twinning was high and lambs born were heavier. Animals licking siella had a glossier coat, which is a sign of good health.

Box 2.9. Federation of women reviving the Aseel chicken breed (ANTHRA, 2008).

The Aseel, a poultry species native to India, is reared under backyard management systems and is a vital source of meat, income and is an important part of Adivasi culture in East Godavari district. The term Âdivâsîs (literally: original inhabitants) is an umbrella term for ethnic and tribal groups believed to be the aboriginal population of India, around 8.2% of the nation's total population, over 84 million people.

Within Adivasi culture, the Aseel chicken is the only resource completely owned and controlled by women; from bird selection to sale. Today this indigenous breed, which has its lineage from the original Red Jungle Fowl, is threatened because of high production losses, infectious diseases and policies promoting non-local breeds.

ANTHRA supported a federation of 1800 Adivasi women across 80 villages to resurrect the Aseel population by building local disease management and feeding strategies, promoting traditional asset sharing to preserve the Aseel biodiversity and lobbying for timely vaccination with government agencies. This good practice shows that it is possible to achieve the following results:

• A remarkable reduction in chick mortality from 70% in 1997 to 25% in 2008 (lowest was 6% in 1999): a threefold increase in income from poultry, comparing the pre-intervention (1998) and actual situation (2008);
• The efficacy of local mobilization, wherein a mass vaccination drive reached out to 12,000 birds in 45 villages;
• Government provided vaccines; trained women vaccinated;
• The value of indigenous Aseel germplasm with average weight of a 2-year-old male bird 3–4 kg and female 2–3 kg;
• The importance of traditional practices with the Aseel having major cultural significance and local market demand – being sold at an average of Rs. 140 per kg – with figures tripling during festival season and fighting cocks priced between Rs. 500 and Rs. 1500 per kg;
• The lack of poultry feed/scavenging material led to a shift in cropping systems (less tobacco, cotton; more food crops); the diversification stimulated a more varied diet and enhanced crop by-products for both poultry and small ruminants.

cause and solve these kinds of challenges, often by finding local solutions to local problems.

One such example is the case of Dutch dairy farmers, who 're-invented' the importance of optimizing farm efficiency through the soil–plant–animal cycle (Fig. 2.4; Box 2.10). Other examples include the work of farmers around the world in the field of organic farming, fair trade and local food chains, often initially against mainstream policies.

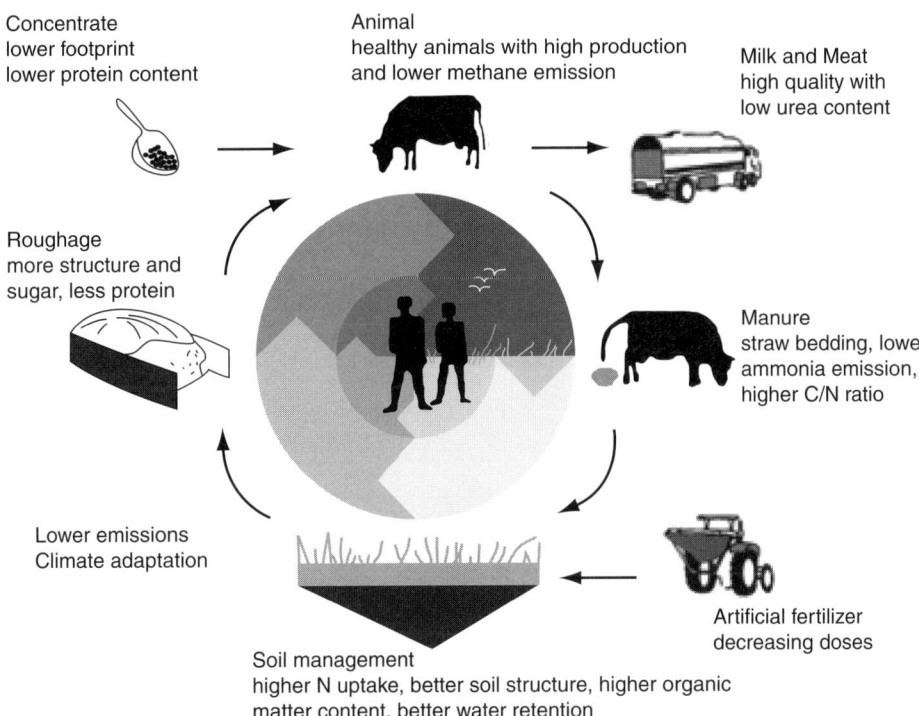

Concentrate
lower footprint
lower protein content

Animal
healthy animals with high production
and lower methane emission

Milk and Meat
high quality with
low urea content

Roughage
more structure and
sugar, less protein

Manure
straw bedding, lower
ammonia emission,
higher C/N ratio

Lower emissions
Climate adaptation

Artificial fertilizer
decreasing doses

Soil management
higher N uptake, better soil structure, higher organic
matter content, better water retention

Fig. 2.4. Soil–plant–animal figure used to explain the ways to optimize nutrient cycles by Dutch farmers.

Box 2.10. Dairy farmers in The Netherlands are re-establishing the soil–plant–animal–manure cycle (Proost and van Weperen, 2006; van't Hooft, 2010). See also Case Study 6, Chapter 12.

Agriculture in The Netherlands only 50 years ago was quite similar to agriculture in developing countries: family farms with a diversity of activities, combining low-input crop production with various species of livestock and numerous other activities. After the hunger of the Second World War, agricultural policies since the 1960s were aimed at 'no more hunger' and aiming at highest yields through specialization, mechanization and market protection (fixed prices): the productivity focus. In this concept, production is measured in litres per day or year, rather than life production. Crop and livestock production became disintegrated, with farms specializing in one or the other.

In the 1960s, the average Dutch dairy cow produced 4000 kg milk per year. In 2001, this was about 8000 kg (Fig. 2.5). After several years, however, the negative side-effects also became clear. The high productivity resulted in low milk prices. Gradually milk producers became dependent on subsidies for a decent income. At farm level, the mineral efficiency (nitrogen and phosphorus) also showed a downward trend, as did farm income and animal life span. The environmental problems led to new policies to handle manure and minerals, which further reduced soil fertility. With declining income and degraded resources, many dairy farmers decided to leave farming. The number of dairy farms has been reduced by nearly 90% since the 1960s (LEI Wageningen UR, 2010).

Today new dairy initiatives are in three main directions: (i) further scale enlargement and intensification; (ii) diversification of farm income, by offering new services (tourism, local markets, care farms); and (iii) increase farm efficiency and total life production on the basis of soil fertility and biodiversity through the cycle approach.

Continued

Box 2.10. Continued.

Fig. 2.5. Dairy cattle being fed high-quality roughage, which improves farm efficiency, manure quality and soil fertility in The Netherlands.

Duurzaam Boer Blijven, which roughly translated means 'continue farming the sustainable way' is an initiative that supports dairy farmers to improve the efficiency of their operations. It builds on initiatives from farmers in the northern province of Friesland who were concerned over problems arising from increasing loss of soil fertility, and the need to comply with new environmental regulations on ammonia and nitrate (Figs 2.6 and 2.7).

The approach encourages dairy farmers to look at their operations in terms of the 'soil–plant–animal–manure' cycle and examine means to reduce environmental impacts through controlling nutrient inputs more carefully. Much of the work is carried out in 'study groups' of farmers that get together to share information and encourage each other to try different activities on their farms.

The two key elements of the approach are effective communications between farmers and with research institutes, and the continuous monitoring of inputs and outputs. Because of this monitoring, the results are consistently being fed back to farmers so that they can see what effect the changes in their operations are having on farm outputs (e.g. milk), the financial situation and the nutrient balance in the soil and water.

Farmers using the approach have significantly reduced their input costs through reductions in fertilizer and concentrates, and through decreased veterinary bills, as fewer animals get sick and they live longer. The approach has also led to improved soil fertility and improved pasture, which results in higher roughage production per hectare of land, but most importantly, more farmers have decided to remain in farming, because of the economic, environmental and social advantages of this system.

Continued

Box 2.10. Continued.

Fig. 2.6. Dutch farmers learn together in study groups, looking at the results of improved soil–plant–animal management on each other's farm.

Fig. 2.7. Soil quality is improved through reduction and adequate use of artificial fertilizer and concentrates, in combination with other soil-improvement practices.

References and Further Reading

Alford, R. and Penney, S. (2006) Preparing to climate proof. The next challenge for Africa's rural poor. Report Send a Cow. The Foundation Series: 'Passing On' learning no. 1. Send a Cow, Bath.

ANTHRA (2008) Unpacking the poor productivity myth. Women resurrecting poultry biodiversity and livelihoods in Andhra Pradesh, India. Good Practice Brief SAGP25, South Asia Pro-Poor Livestock Policy Programme, FAO/NDDB, Rome.

FAO (2006) *Livestock's Long Shadow – Environmental Issues and Options*. FAO, Rome.

FAO (2009) *Food Security and Agricultural Mitigation in Developing Countries: Options for Capturing Synergies*. FAO, Rome.

FAO (2010a) Climate-smart agriculture policies, practices and financing for food security, adaptation and mitigation. Technical input for the Hague Conference on Agriculture, Food Security and Climate Change, 31 October–5 November. FAO, Rome.

FAO (2010b) Greenhouse gas emissions from the dairy sector – a life cycle assessment. Report prepared by Food and Agriculture Organization of the United Nations – Animal Health Division. FAO, Rome.

Geerlings, E. (2007) Highly pathogenic avian influenza: a rapid assessment of the socio-economic impact on vulnerable households in Egypt. FAO/WFP Joint Project Report. Rome.

Hooft, K. van't (2004) *Gracias a Los Animales – Crianza Pecuaria Familiar en America Latina, Con Casos de los Valles y el Altiplano de Bolivia (Thanks to the Animals – Family Level Livestock Keeping in Latin America with Case Studies from the Bolivian Valleys and Highlands)*. Plural Publishers, La Paz.

Hooft, K. van't (2009) Livestock friend or foe, the need to look at different production systems in the debate about livestock & climate change. Available at Endogenous Livestock Development Network. http://www.eldev.net/

Hooft, K. van't (2010) The role of ethnoveterinary practices in post-modern agriculture – examples from The Netherlands. In: *Proceedings of the International Conference on Ethnoveterinary Practices*, Thanjavur, Tamil Nadu, 4–6 January.

IPCC (2007) Contribution to the Fourth Assessment, Report of Inter-Governmental Panel on Climate Change. Working group II: Adaptation and Vulnerability, and Working group III Mitigation of Climate Change. IPCC, Geneva.

Karbo, N. (1999) Natural mineral licks to enhance livestock growth. CSIR Animal Research Institute, Tamale, Ghana. In: *Appropriate Technology*, volume 34, no. 1. Livestock in Development, Crewkerne.

Kristjanson, P., Waters-Bayer, A., Johnson, N., Tipilda, A., Njuki, J., Baltenweck, I., Grace, D. and MacMillan, S. (2010) Livestock and women's livelihoods: a review of the recent evidence. Discussion Paper No. 20. ILRI, Nairobi.

LEI Wageningen UR (2010) Mineralenmanagement en economie op melkveebedrijven – gegevens uit de praktijk (Mineral management and economics on dairy farms – data from practice). LEI Wageningen Research Sector & Entrepreneurship brochure no. 09-066.

Letty, B.A. and Waters-Bayer, A. (2009) Recognising local innovation in livestock-keeping – a path to empowering women. Paper in WCAP.

Livestock in Development (1999) Livestock in poverty-focused development. http://www.smallstock.info/reference/LID/livestock.pdf. Livestock in Development, Crewkerne.

Mathias, E. and Mundy, P. (2008) Endogenous livestock development: strengthening local initiatives and using local resources sustainably. League for Pastoral Peoples and Endogenous Livestock Development, Ober-Ramstadt/ELD Network, Leusden.

Perezgrovas, R. (2006) Direct involvement of indigenous women in sheep improvement research in Chiapas, México. Prize winning poster in German Tropentag 2006, Bonn: Prosperity and Poverty in a Globalized World – Challenges for Agricultural Research.

Proost, J. and van Weperen, W. (2006) Creating space for change: farmers' learning groups in The Netherlands. In: *Compas Magazine* July.

3

Methodologies, Organizations and Networks in Endogenous Livestock Development

─────────────

Learning Objectives: Understanding
- The role of livestock in integrated agricultural systems
- The endogenous development concept
- Endogenous livestock development (ELD)
- Methodologies to support ELD
- Organizations and networks engaged in ELD initiatives

Role of Livestock as Part of Integrated Agricultural Systems

The word agriculture is comprised of two parts: *agri* and *culture*. This highlights the fact that in many societies, agriculture does not simply imply the production of a crop or livestock; it is actually their way of life. This is still expressed in many rural communities today. In many developing countries, up to 95% of families receive their livelihoods from agriculture every day (Fig. 3.1). Within these societies, meat, milk and eggs are not neatly packaged in cardboard or plastic containers, and the role of livestock is not debated.

Livestock are an integrated part of the agricultural systems that stand at the basis of the farms and family life in many parts of the world. In most rural communities in Asia, Latin America, Eastern Europe and Africa, livestock are essential for draft power, organic fertilizer and transportation. They are a source of nutritional protein, wool, leather and fibres.

Raising animals is also a form of security against climatic and economic risks, and a means of accumulating and maintaining financial reserves. In this sense in many cultures, animals are not only seen as a cash investment, but rather as 'wealth' in a more cultural sense. Besides money, this traditional concept of wealth includes family, land, respect, knowledge and skills, happiness, acceptability and satisfaction (van't Hooft *et al.*, 2008).

Animals also play a significant role in cultural and spiritual life, especially linked to cultural identity (Fig. 3.2). There are countless examples of this. Many pastoral people, such as the Fulani in western Africa, have linked their myth of origin to taking care of their animals. This aspect of identity is also expressed into animals as 'totem symbols'. In many cultures, no ritual or ceremony can take place without the ritual slaughter of an animal; animals are presented as a gift during marriage or maturity rituals. Women in parts of south Asia can

©K.E. van't Hooft, T.S. Wollen and D.P. Bhandari 2012. *Sustainable Livestock Management for Poverty Alleviation and Food Security* (K. van't Hooft, T. Wollen and D.P. Bhandari)

> **Box 3.1.** Farmers' voices about the diverse and complex role of animals (van't Hooft *et al.*, 2008).
>
> Livestock are part of our lives, our survival depends upon them, without them we have no life. They provide milk, money, manure. They pay our hospital bills and education. They are like brothers and sisters to me. We use livestock to create good relationships with our family members, or to support family members that have been afflicted by a disaster. Indirectly the manure has helped us to develop a good relationship with non-pastoralist communities, because it has improved the fertility of their lands.
>
> El Haji Eggi Sule, Mbororo pastoralist, ethno-vet healer in Ntam Village, Bui division, Cameroon
>
> We use animals for celebrating deaths, births and marriages. Also for food and manure. Since the Holstein breed is new to our society and their prices are higher, I might sell a bull and buy a local breed and reserve them for these celebrations, like the Goudali. Animals and crops are a two-way traffic. Crops provide fodder for the animals and manure supplies nutrients for the crops.
>
> Stephan Ndonwi, dairy farmer, Akum village, Mezam, Cameroon

Fig. 3.1. Livestock are an integral part of the agricultural systems that stand as the basis of the farms and family life in many parts of the world.

Fig. 3.2. In many cultures cattle are considered holy. Credit: Compas.

have animals at their own possession, which helps them to buy items they use to beautify themselves. While the cultural aspect of human–livestock relationships is often overlooked, it is vital to the farmers in rural and non-industrialized countries, where poverty remains an alarming problem.

Endogenous Development

The word endogenous means 'growing from within'. Endogenous development as an approach evolved out of the school of action research and participatory approaches in agriculture and natural resource management in the late 1980s. During the course of the 1990s, the importance of participatory approaches

and of integrating local knowledge into development interventions became broadly recognized. However, in practice, many of the approaches that had been developed experienced difficulties in overcoming an implicit 'materialistic' bias, where the focus was on physical and economic development.

Endogenous development seeks to overcome a materialistic bias by making peoples' worldviews and their livelihood strategies the starting point for development. Many livelihood strategies reflect notions of sustainable development as a balance between material, social and spiritual well-being. These three dimensions are seen as inseparable (Fig. 3.3).

Endogenous development is already present in local communities. It is reflected in the communities' capacities for self-

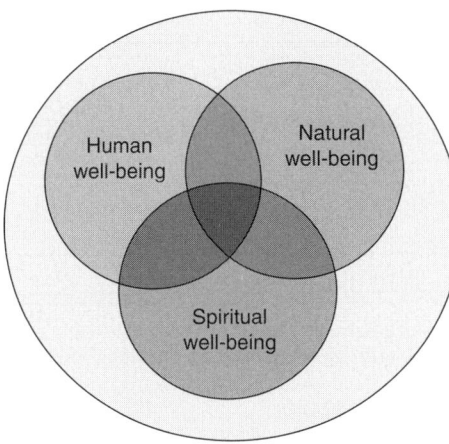

Fig. 3.3. The aim of supporting endogenous development is to empower local communities to take control over their own development process in order to achieve their material, social and spiritual well-being aspirations. Credit: Compas.

determination. Outside organizations can support and strengthen the endogenous development process within communities. The aim of supporting endogenous development is to empower local communities to take control over their own development process in order to achieve their material, social and spiritual well-being aspirations (Lammerlink and Otterloo-Butler, 2010).

In endogenous (livestock) development it is recognized that local identity, or the worldview of the people involved, underpins their use of these local resources. In these worldviews – and thus in the use of local resources – a balance is sought between three spheres of life: the human world, the natural world and the spiritual world. This is reflected in human–animal relations, and how livestock systems are built (van't Hooft *et al.*, 2008).

Endogenous Livestock Development

Endogenous livestock development (ELD) seeks to enhance livestock keepers' own development efforts. This includes both the owners of animals and people that depend on animals and their products in other ways. ELD stands for supporting husbandry systems that are based on livestock keepers' own innovative strategies, knowledge and resources, as well as their perception of well-being and improvement. It thus indicates the need to work in a 'people-centred' rather than an 'animal-productivity centred' way. Therefore, it is also known as people-centred livestock development.

The ELD methodology includes supporting local innovation and understanding people's own worldviews, including their criteria for development, learning, experimenting and communication. It also addresses the 'why' of local practices, knowledge and existing leadership structures, and stimulates openness towards cultural practices.

Putting livestock keepers at the centre of their own development requires a basic rethink of how livestock are produced. Outsiders can facilitate that process. They can help marginalized and poor livestock keepers gain recognition and support for their initiatives. This can be done through training, research, advisory services, advocacy and networking. That is what ELD is all about.

Mary Sirri also stresses the effect of ELD on the relations between farmers and supporting organizations:

Before, Heifer used to do everything, but now we can make an action plan by ourselves. We are sharing knowledge and experiences with each other. Heifer has made us understand that we should not sit and depend only on them. We should come out with our own plan, and if Heifer comes out with something that we don't like, we should express ourselves. This change came through the workshop. We were just depending on them. Before, we always dewormed the cows every three months. Now, the trained farmers should deworm them. Let every cooperative develop an action plan for deworming and execute it. I think more critically about my farming activities now, thinking more ahead. I think Heifer is happy with the change in us; we are now more active and critical.

Box 3.2. What is ELD?

ELD is about enhancing the capacity of livestock keepers to solve their own problems. Based on their own culture and worldview, endogenous development farmers are supported to develop technologies and skills that broaden the options available to them. In this way, a balanced interface is promoted between tradition and modernity.

Box 3.3. Farmers' voice about the ELD (van't Hooft *et al.*, 2008).

Endogenous livestock development means taking action yourself, starting from your own doorstep. I can start with what is within my reach before going ahead to ask from somebody. Before doing that I should think of the people who are around me, our ancestors, our children and what will become of us in the future. ELD is helping people to stand on their own feet and work together.

Mary Sirri, dairy farmer, Akum village, Mezam Division, Cameroon (Fig. 3.4)

Fig. 3.4. Mary Sirri, a dairy farmer from the Akum village in the Mezam Division of Cameroon, talking about changes brought about by working with NGOs.

According to Isaac Gabesin, government veterinary nurse, Cameroon:

> endogenous development is now a blend of the two, i.e. you have a mixture of both internal and external development approaches and then you blend the two and see what is really best for the society. So we don't over-focus ourselves on one and leave the other. We work with the people and then see which best suits the people.
>
> (van't Hooft *et al.*, 2008)

Supporting Endogenous Livestock Development – in Practice

ELD activities, such as those mentioned in the last section of Chapter 2, are from different parts of the world but they have several things in common:

- They are all driven by livestock keepers and/or family farms – not by outsiders. That means that, unlike outside interventions, they address the problems that livestock keepers themselves face and understand.
- They are environmentally friendly. They use local resources, avoid pollution and conserve the environment.
- They make economic sense. They use relatively few costly external inputs, generate competitive levels of output and are profitable for the farmers involved.
- They are based on joint learning. They include a combination of local and

outside expertise. Outside expertise and resources are tested by farming families, and in the process adapted to local conditions and needs.

These and similar local initiatives deserve special recognition and support. Smallholder livestock keepers themselves should be included in decision making on issues that affect them. Those issues include the use and management of natural resources, access to land, land credit and markets, intellectual property, research and trade-priorities, and protection of rural environment (Mathias and Mundy, 2008).

Over the past few decades, many organizations throughout the world have developed people-centred livestock development approaches. These include participatory learning and action, partici-patory innovation development and endogenous development. In the field of livestock, these are people-centred rather than animal-productivity-centred approaches, directed at poverty allevia-tion through participatory, low-input livestock development. Some focus on ethno-veterinary practices, others on fam-ily poultry, smallholder farms, integrated animal agriculture, farmer field schools for innovation and diffusion of approp-riate technology, pastoralism, training community animal health workers or supporting local innovations in livestock.

The Need to Maintain Livestock Biodiversity

Local breeds are the result of centuries of pur-posive selection, socio-cultural influences and traditional knowledge combined with natural selection. Their value is often neglected just because of perceived lower productivity. Local breeds of animals are maintained by smallholders and pastoralist societies.

Many breeds are threatened because agriculture has changed. Modern food pro-duction now favours the use of a few highly specialized breeds selected for maximum output in a controlled environ-ment. In the name of producing more

---

Box 3.4. NGO deputy director's voice about the ELD (van't Hooft *et al.*, 2008).

My attitude towards people changed. I knew that people in the communities already know things, but learn-ing that others have already been thinking like this [*within the ELD network*] made it come with full force. It gave a greater push to the ideas that were already in my mind. We have to design things with people and not on our own. Before planning was done in the office, but since the past 3–4 years things are changing and the workshop gave us a push towards enhancing the participatory process. We're realizing that farmers have a lot of knowledge and when projects are planned this is now taken into consideration.

Janet Akob, former deputy country director Heifer Cameroon and coordinator gender & HIV/AIDS programme (Fig. 3.5).

...when these persons realize that, well they know something, then they can open up.

Fig. 3.5. Janet Akob (left), former deputy country director Heifer Cameroon: 'It was an eye-opener to know that improving productivity and reducing poverty does not only mean bringing things from outside. It was a strong turning point for me to realize that we can enhance what people already have. This gives a new dimension to our work.' Credit: Ellen Geerlings.

Fig. 3.6. Local breeds and strains of poultry are more capable of scavenging for food and protecting themselves from predators. Credit: Ellen Geerlings.

animal source protein, local breeds of livestock are disregarded and many are prone to extinction.

Meanwhile, there is a growing recognition of the need to protect genetic diversity in livestock species through the *in situ* conservation and promotion of endangered breeds within the communities (Fig. 3.6). Rare or heritage breeds are kept for several reasons. One is the genetic preservation aspect. These breeds are part of the ecosystem where they were developed and represent a unique piece of earth's biodiversity. Each breed carries a unique capability into the survival of the species and provides a solution to some challenge that may threaten its survivability in the future. A second reason for maintaining local breeds is that of adaptation to the local environments or local forages; they have the capability to do well on fewer external inputs.

Loss of any one local breed impoverishes agriculture and long-term viability of the species. Mankind has inherited a rich variety of livestock breeds. For the sake of future generations, we must work together to safeguard these treasures. Not only do they evoke our past, they are also an important resource for our future (LPP, LIFE network, IUCN-WISP and FAO, 2010).

Passing on the Gift and Integrated Animal Agriculture

Heifer International aims to provide quality livestock, training and related support to men, women and youth in order to assist with food security and to improve livelihoods. The approach includes the overall development of a community of families in resource constraints. The activities are implemented around strong local community groups. The community group decides on the initial receivers of the animals. The first female offspring is then passed on to another trained member of the community group. The families are supported to put integrated animal agriculture into practice. A recent study of this system in Uganda by Send a Cow (Alford and Penney, 2006) has shown the positive effects of this system, not only in terms of food security and income, but also in carbon sequestration, soil fertility and increase in fodder trees.

Action Research Methodology

ELD takes the best of both local and global. It supports producers to build on what they already own and do, and takes advantage of their indigenous knowledge. It draws on modern technology where appropriate. This becomes clear, for example, in the action research methodology (Box 3.6).

Ethno-veterinary Medicine – Supporting Indigenous Knowledge of Livestock

Ethno-veterinary medicine deals with people's knowledge, skills, methods, practices and beliefs about the care of their animals. Ethno-veterinary knowledge is acquired through practical experience and has

---

Box 3.5. Passing on the Gift in Vietnam (Matthews, 2007).

Heifer International encourages livestock recipients to establish an integrated farming system where crops and livestock complement each other. One example of the benefits of this system is the farm of Nguyen Buu Chau in Vietnam, developed with support from Heifer International.

Nguyen's farm exemplifies how the integrated farming system works. A barn has been constructed to help manage the cows he received from Heifer International. A few steps away are the kitchen with a vegetable garden and the farm's crops. Beside a canal in the back, Nguyen keeps low, brick worm bins where chickens and ducks congregate. Cows are fed concentrates and leftovers from the kitchen, as well as elephant grass on the edge of the rice field. His children help to cut and feed the grasses as a part of their daily activities. The cow manure fertilizes the crop fields and kitchen garden. The manure also helps to recycle the crop nutrients by promoting healthy microbial action in the soil.

Heifer International-Vietnam also helped Nguyen install a biogas unit below ground, which feeds methane from cow manure to the kitchen stove. The biogas unit is the primary method of cooking for the household, but provides more for the family than just a cooking source. The leftover sludge from the biogas is used as fertilizer to increase rice and corn production. Not all of the cow manure goes into the bio-gas unit. An above-ground compost heap is home to thousands of earthworms that turn organic waste into worm castings of concentrated nutrients, which become a natural plant fertilizer. Once separated from the castings, the worms make a high-protein feed for the chickens and ducks. The poultry in turn provide eggs, meat and insect control, and the manure from these birds is rich in nitrogen and phosphorus, further fertilizing the garden.

---

traditionally been passed down orally from generation to generation. 'Ethno-veterinary medicine' and 'ethno-animal science' focus on livestock keepers' approaches to animal health and production. They cover herbal medicines and other natural substances to treat a wide range of ailments, housing, feed, reproduction and many other issues. It is necessary to look at what these practices are and how they work as well as why people use them. This implies understanding local people's cultural background and worldviews (FRLHT and Tanuvas University, 2010; Agromisa Foundation and CTA, 2007). Further information can be obtained from the following sources: www.ethnovetweb. com, www.anthra.org, www.compasnet.org, www.frlht.org, www.iaim.edu.in.

Participatory Innovation Development

Livestock keepers often have unique and useful ideas on how to improve their own production systems. They have numerous innovative ideas and practices, which can enrich the research and development agenda.

Research and development workers can stimulate the exchange on these innovations, include them into research agendas, and build on this knowledge. In this way, possibilities can be explored jointly and new ideas can be tested. Local and external sources of knowledge can be combined to create solutions that fit local situations (Sharad, 2006). Further information can be obtained from: www. prolinnova.net.

Handing Over the Stick

Handing Over the Stick is a hands-on workshop that engages community members and teaches participants concrete approaches to planning, decision making and assessment.

Outside facilitators use participatory approaches to encourage livestock keepers to define their problems from their own points of view, and to realize their potential to solve them. Then they 'hand over the stick', enabling the livestock keepers to take control over the development process. His/her approach is very flexible and can imply a whole

Box 3.6. Action research on beekeeping (van't Hooft *et al.*, 2008).

Farmer Joseph Mboussi from Cameroon has been working with bees for a long time.

In the past, when I found a hive within a tree, I would burn it in order to get to the honey. This resulted in a very wasteful system, while the honey that resulted from it was of low quality. Then I learned about beekeeping from an organization, and that it is possible to have beekeeping husbandry, similar to sheep keeping. I started to use boxes with small panels adapted to the size, which I can take out in order to harvest the honey. I felt that it was better to do it in this way, also for the environment. I also became more aware of the risks of bushfires, the need for hygiene and how to control a moth that attacks the bees. Within my community, we have organized a group to enhance beekeeping. We felt that our production was still very low, and that it would be necessary to have more hives, in order to have more bees, get more honey and also make use of the by-products, like pollen, propolis and wax. We place empty boxes near trees with wild beehives and wait for the bees to colonize them. The three major problems we have are: there are not enough bee colonies; we need more hives; we lack pollinating plants to provide the food for the bees.

After this training, the group started to discuss ways to solve these problems. Gradually a shift was made from 'looking for support from outside' to *looking for things to try,* which they could do themselves while looking for complementary outside support. It was concluded that their development partner should not simply provide them with hives, but rather look at the beekeeping system the people have devised here, and experiment jointly with the farmers on the questions they have. The outcome of this could then also be used in other areas. So they began to do action research together to find the best hives and how to increase the number of bee colonies. Criteria set for the research included finding a balance between: (i) quality versus the quantity of the honey; (ii) the bees' requirements and preferences versus the easiness of handling the bees; and (iii) using the colonizing versus the non-colonizing (wild) method.

The training sparked much farmer exchange. For example, one farmer showed his experience with filtering the honey, using a bucket and a very clean piece of cloth. Other contributions included ways to prevent termites, various modern and traditional ways of extracting honey, and ways to enhance the number of pollinating trees. This discussion made us aware of the importance of changing the concepts of 'training' (in conventional, top-down ways) into more joint action research between farmers and NGO fieldworkers. In this way, the farmers can be in the driving seat of the research. Supporting NGOs can provide the farmers with essential information they have no access to, and facilitate certain essential inputs when the need arises within the process of action research.

Fig. 3.7. Small-scale, backyard beekeeping – such as here in Haiti – is becoming more popular and profitable.

range of methods that can be adapted as needed. For more information, visit www. anthra.org.

Training Community Animal Health Workers

Many rural communities and livestock holders have many animals but few veterinarians. Even more easily accessible sedentary livestock producers find it hard to obtain the services they can afford: veterinary care, breeding stock, credit, and so on. One way to overcome this is to involve local people themselves in providing these services. Examples include training 'para-veterinarians' or community animal health workers to deal with simple animal health and nutrition problems, promote local efforts to conserve breeds and develop animal breeding programmes, all in the support of local institutions and self-help groups (Fig. 3.8).

A community animal health worker is a specially trained local community member who helps farmers and community groups to raise healthy animals to maximize their benefits. The community animal health workers are paid for their work in cash or in kind, by the local community or by the farmer whose animal receives

treatment. The primary role of the community animal health worker is to reduce mortality and increase productivity in local livestock through the increased access to affordable, basic, animal health services (Bhandari, 2010). Most community animal health workers are using modern drugs; others are trained to combine modern drugs with medicinal plants and other ethno-veterinary practices. Many organizations are involved in these trainings, such as: www.heifer.org, www. farmafrica.org.uk, www.ahtcs.org.np, www. vsfe.org, www.anthra.org.

Supporting Livestock Keepers' Rights through Bio-cultural Community Protocols

Livestock keepers rely not only on their animals; they also need grazing land, water, markets, veterinary care and information. National governments and international conventions decide on who has access to these resources, but livestock keepers, especially the poor and unorganized livestock keepers, are usually frozen out of negotiations. NGOs have been helping them to defend their rights and livelihoods. Prominent issues include the right to maintain their own animal breeds, a practice that is increasingly being

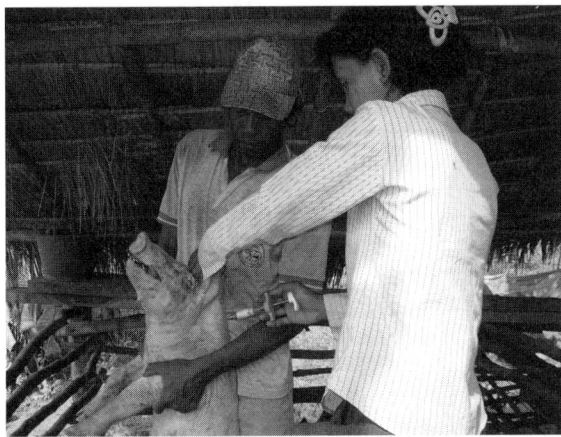

Fig. 3.8. Community animal health worker with the Animal Health Training and Consultancy Service (AHTCS) in Nepal.

Fig. 3.9. Bio-cultural Community Protocol of the Raika, an indigenous pastoral community living in Rajasthan, India. They have grazed their livestock for over 700 years on communal land and forests. Their traditional culture, values and livelihoods are inextricably linked with their animals and surrounding ecosystems. Credit: Ilse Koehler-Rollefson.

restricted. It also implies the use of traditional grazing lands, and maintenance of a pastoral lifestyle. Other negotiations affecting livestock keepers concern human rights, indigenous peoples, indigenous knowledge, trade, intellectual property rights (for example in the use of medicinal plants), and the right to food (United Nations Environment Programme, 2009).

A BCP is a protocol that is developed after an indigenous or local community undertakes (Fig. 3.9) a consultative process to outline their core values and customary laws relating to their traditional knowledge and resources, such as their local breed. On this basis, they provide clear terms and conditions for outsiders to regulate access to these resources. For more information, see www.naturaljustice. org.za, www.pastoralpeoples.org, www. compasnet.org.

Livestock Emergency Guidelines and Standards

The Livestock Emergency Guidelines and Standards (LEGS) have been developed as a set of international guidelines and standards for the design, implementation and assessment of livestock interventions to assist people affected by humanitarian crises. Tufts/Feinstein International Center (FIC) faculty have been instrumental in leading the coalition to develop these standards, which were published in early 2009 and are available as a free download on the LEGS website (http://www.livestock-emergency.net).

The LEGS process grew out of recognition that livestock are crucial livelihood assets for people throughout the world, and livestock interventions are often a feature of relief responses. Yet, there were no widely available guidelines to assist donors, programme managers or technical experts in the design or implementation of livestock interventions in disasters. LEGS recognizes that climatic trends are causing more frequent and varied humanitarian crises, particularly affecting communities who rely heavily on livestock. In 2009 and 2010, the FIC secured funding from the UK Department for International Development (DFID) to support LEGS awareness raising and training activities in Africa and Asia (LEGS, 2010).

Labelling for Ecological Animal Husbandry

Ecological animal husbandry is based on a vision of a sustainable society and a more sensible way of doing agriculture: animals should be part of an agricultural system that is environmentally sound, and it should be animal and human friendly as well. It calls for a holistic or integrated approach, which considers the whole system rather than only optimizing its parts.

Livestock production in developing countries often uses few outside inputs and is ecologically sound. At the same time, it is neither connected with nor can meet the standards of organic animal production in the developed world. In this way, they are cut out of the often booming eco-market.

More universal standards are needed that better fit developing country niche market situations. One solution might be a product label for pastoralists, a *range-fed* label that would distinguish meat coming from pastoralists' animals from large-scale producers (DARCOF, 2000).

Promoting Local Breeds through Marketing of their Products

Finding and promoting niche markets for products of local breeds is a possible way of ensuring their survival and enabling the people who keep them to earn more from their existing lifestyle. There are numerous examples of such efforts, eight of which are described in a recent publication by League of Pastoral Peoples (LPP), LIFE Network, IUCN-WISP and FAO (2010). The cases deal with products like wool, meat and milk from various animal species, and include four main types of intervention: improving animal production, processing, organization and building a value chain; www.pastoralpeoples. org. (See also Chapter 11.)

Additional National and International Networks

ELD network

The ELD network brings together various approaches and organizations working with livestock-related participatory development initiatives. The network also brings together around 450 professionals working in livestock development by means of an active e-mail exchange. For more information and to link up with the ELDev list, go to http://www.eldev.net/

Community of Practice for Pro-Poor Livestock Development

The Community of Practice for Pro-Poor Livestock Development (CoP-PPLD) is an online sharing network for practitioners, managers, researchers and other actors involved in pro-poor livestock development that want to exchange experiences, innovative approaches, best/next practices and other knowledge (including tacit) for the CoP-PPLD's mutual learning. The shared goal is to learn from and give a voice to the livestock community regarding a wide range of issues affecting the poor livestock keepers today, contributing thus to livestock development as an instrument for poverty reduction. http://www.cop-ppld.net/

Domestic Animal Diversity Information System

DAD-IS is the Domestic Animal Diversity Information System hosted by FAO. It is a communication and information tool for implementing strategies for the management of animal genetic resources (AnGR). It provides the user with searchable databases of breed-related information and images, management tools, and a library of references, links and contacts of Regional and National Coordinators for the Management of Animal Genetic Resources. It provides countries with a secure means to control the entry, updating and accessing of their national data. http://dad.fao.org/

Heifer International – Passing on the Gift

Heifer International is a global non-profit organization with a proven solution to ending hunger and poverty in a sustainable way. Heifer helps empower millions of families to lift themselves out of poverty and hunger to self-reliance through gifts of livestock, seeds and trees, and extensive training, which provide a multiplying source of food and income. Since 1944, the total number of families assisted directly and indirectly amounts to more than 70.5 million men, women and children with project communities in Africa, throughout the Americas, Asia/South Pacific and Central and Eastern Europe. www.heifer.org.

Vétérinaires Sans Frontières Europe

Vétérinaires Sans Frontières (VSF) Europe is an international non-profit organization that aims to improve the living conditions of the most vulnerable societies by supporting smallholder farmers, pastoralists and rural livelihoods. VSF Europe is a network of ten member organizations operating in over 40 countries worldwide. The main programmes include supporting community-based animal health programmes, support to local veterinary services, training animal health workers, promotion of traditional veterinary knowledge, support to market access, emergency relief, promotion of food sovereignty and income generating activities. www.vsfe.org.

Bureau for Exchange and Distribution of Information on Mini-livestock

BEDIM is the Bureau for Exchange and Distribution of Information on Mini-livestock, including rodents, guinea pigs, frogs, snails, worms and insects. This is a non-profit international organization devoted to identification, processing and diffusion of information and data concerning animal species related to mini-livestock and their products. www.bedim.org.

Smallstock in Development

Smallstock in Development is an Internet toolbox (2006) developed by NR International, managers of the Livestock Production Programme (LPP) funded by the UK's DFID.

Livestock in general, and smallstock in particular (including sheep, goats and poultry), have an important role to play in enhancing the livelihoods of the poor.

In poor households, these animals are often kept under scavenging conditions with little or no attention paid to supplementing feed inputs, or to disease control and housing. At the same time, these animals provide products for cash sale when a need arises, and provide the household with much needed protein.

The Smallstock in Development toolbox focuses on the role and importance of smallstock in development and poverty reduction. The toolbox also aims to provide a range of practical information and descriptions of techniques or 'tools' to assist in increasing the efficiency of operations of smallholders and/or the productivity of their animals. www.smallstock.info

American Livestock Breeds Conservancy

Founded in 1977, the American Livestock Breeds Conservancy (ALBC) is the pioneer organization in the USA working to conserve historic breeds and genetic diversity in livestock. The ALBC mission is stated: 'Ensuring the future of agriculture through genetic conservation and the promotion of endangered breeds of livestock and poultry'. ALBC is a non-profit membership organization working to protect over 180 breeds of livestock and poultry from extinction. Included are asses, cattle, goats, horses, sheep, pigs, rabbits, chickens, ducks, geese and turkeys. www.albc-usa.org.

Society of Animal, Veterinary and Environmental Scientists

The Society of Animal, Veterinary and Environmental Scientists (SAVES) is an initiative founded as a Pakistani society, and has now become an international network aimed at the conservation of animal genetic resources and indigenous knowledge through the strengthening of pastoral peoples. www.researchgate.net/group/SAVES_society_of_Animal_Vet_and_Environmental_Scientists/

References and Further Reading

Agromisa Foundation and CTA (2007) *Ethnoveterinary Medicine*. Agrodoc. Agromisa Foundation, Wageningen.

Alford, R. and Penney, S. (2006) Preparing to climate proof. The next challenge for Africa's rural poor. Report Send a Cow. The Foundation Series: 'Passing On' learning no. 1. Send a Cow, Bath, UK.

Bhandari, D.P. (2010) *Community Animal Health Worker Manual*. Heifer International, Little Rock, Arkansas.

Compas (2010) Community Wellbeing Assessment. Policy brief. Compas network of ETC Foundation. Compas partner Agruco, La Paz. www.compasnet.org

DARCOF (2000) Ecological Animal Husbandry in the Nordic Countries. DARCOF Report No 2, 2000. Danish Centre for Organic Farming, Tjele.

FRLHT and Tanuvas University (2010) Report on International Conference on Ethnoveterinary Practices. Mainstreaming Traditional Wisdom on Livestock Keeping and Herbal Medicine for Sustainable Rural Livelihood across Continents, Thanjavur, Tamil Nadu, 4–6 January.

Hooft, K. van't, Millar D., Geerlings, E. and Django S. (2008) *Endogenous Livestock Development in Cameroon – Exploring the Potential of Local Initiatives for Livestock Development*. Agromisa Foundation Wageningen, The Netherlands.

Lammerlink, M. and Otterloo-Butler, S. (2010) *Seeking Strength from Within – The Quest for a Methodology on Endogenous Development*. Compas network of ETC Foundation.

LEGS (2010) Livestock Emergency Guidelines and Standards. Livelihoods-based livestock interventions in disasters. Practical Action, Rugby, UK.

LPP, LIFE network, IUCN-WISP and FAO (2010) Adding Value to Livestock Diversity. Marketing to promote local breeds and improve livelihoods. FAO Animal Production and Health Paper, no. 168. FAO, Rome.

Matthews, J. (2007) Integrated farming in Vietnam. In: World Ark, July–Aug. Heifer International, Little Rock, Arkansas.

Mathias, E. and Mundy, P. (2008) Endogenous livestock development: strengthening local initiatives and using local resources sustainably. League for Pastoral Peoples and Endogenous Livestock Development and the ELD network.

Sharad, R. (2006) Guidelines to Participatory Innovation Development. PROLINNOVA Nepal Programme.

United Nations Environment Programme (2009) Bio-cultural Community Protocols: A Community Approach to Ensuring the Integrity of Environmental Law and Policy.

4

Differentiating Four Livestock Production Systems

Learning Objectives: Understanding
- The four major livestock production systems
- The characteristics of the four major livestock production systems in developing countries
- The characteristics of the four major livestock production systems in developed countries – with The Netherlands as an example

Four Basic Livestock Production Systems

This chapter describes the four basic livestock-keeping systems (Fig. 4.1) that exist within both developing and developed countries. With a better understanding of each of these, we can learn to what extent changes can be made to improve each of them. Each of these four systems has its own specific objectives, potentials, limitations and 'right of being'. At the same time, these livestock-keeping systems influence environment and livelihoods in different ways – and can be optimized in a sustainable way, taking into account economic, social and environmental considerations.

These systems are not divided according to rich or poor countries. Even in the wealthiest countries and regions of the world, there are farm families of limited means and whose farming operations are low-input. Likewise, even in the poorest of nations there are rich landlords with large-scale livestock operations, who can feed their animals better than the poor can care for their families.

There is, however, a marked difference in the relevance of each of the systems between developing and more developed countries. In most developed countries, high-input and specialized systems cover most of the farm acreage and animal numbers, and provide most of the milk, meat, eggs and other by-products for the urban markets.

In most developing countries, the low-input systems – smallholder farmers and pastoralists – cover most land, engage most people and provide most of the milk, meat and eggs required.

Livestock-keeping Systems in Developing Countries

Though there are differences between each country and region, most livestock-dependent poor people can be found in the smallholder farming and pastoralism systems. Besides

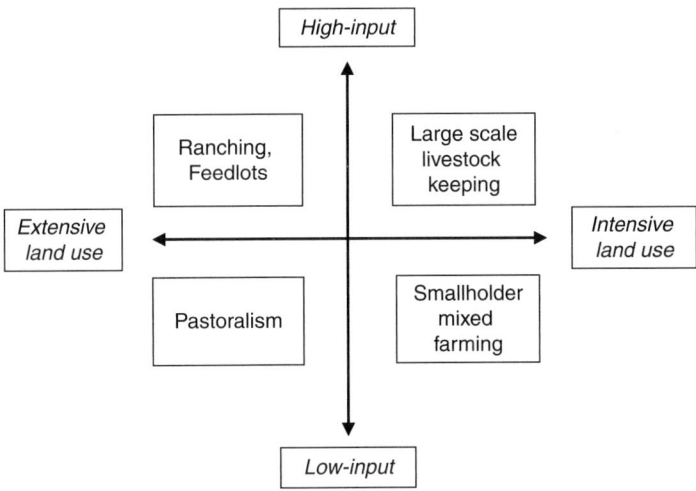

Fig. 4.1. The four major livestock-keeping systems (van't Hooft *et al.*, 2008). On the horizontal axis they differ according to the intensity of land use, varying from extensive land use (left) to intensive land use (right). On the vertical axis, they differ according to the level of inputs within the system, varying from low-input (bottom) to high-input (top).

producing milk, meat and eggs, their animals also provide valuable manure, draft power, banking services and status. A combination of the different systems is often found within one household or farm, e.g. when an intensive dairy cattle farmer also keeps chickens and pigs on a low-input basis or when a smallholder farmer keeps sheep on a semi-pastoralist basis. Meanwhile, in most developing countries, the high-input specialized systems are growing in size and numbers, with large-scale animal farming especially around urban centres.

Intensive land use and low-input: smallholder mixed farming

The smallholder farming system (Fig. 4.2) with intensive labour practices often involves the whole family. Livestock are combined with low-input, rain-fed crop production and usually some non-farm income. This system requires diverse inputs and labour, and can be found in marginalized and isolated areas as well as periurban locations. Farm families use most inputs from local or natural resources. Here, indigenous knowledge, local practices, local breeds or crossbreeds are prevalent. Animals have various roles and production purposes, with numerous outputs: milk, meat, manure, draft power, transport, urine – to name but a few. There is a major role for women and often the need for involvement of children. Output is aimed at domestic consumption, local markets and non-monetary exchange. Constraints on land use are often related to population growth.

There is a great potential for poverty alleviation and maintenance of environmental systems by optimizing the smallholder mixed farming systems. At the same time, relatively limited attention is paid to it by policy makers, research or extension education. Improvements can happen when there is effective support to social networks, culturally sensitive approaches, adapted technical support and financial mechanisms designed for low-income borrowers.

Extensive land use and low-input: (agro-)pastoralism

Pastoralists are known for extensive, low-input livestock-keeping systems, often in ecologically marginalized dryland areas (Fig. 4.3). Local knowledge and local practices are of extreme importance with diversity of breeds, feeds and roles of animals. Markets are usually local or national. There are constraints related to changes in land-use, climate change as well as

conflicts with settled farmers. Government policies are often aimed at re-settling rather than supporting these pastoralist societies. This system has a direct role in poverty alleviation and maintenance of fragile environmental systems. It generally receives little attention from research and education.

There are various organizations that aim to improve the viability of (agro-)pastoral livelihoods even in the midst of severe challenges from changes in climate, land use and markets. Pastoralists are devising new practices that enhance their business outcome (see Chapter 11). Communication technology – such as mobile phones – is providing additional benefits through greater access to market and meteorological information. To know when the price of cattle, goats, sheep or camels is up or down, combined with weather forecast information, gives the agro-pastoralist the ability to control the time to move into or hold back from the market to sell their stock.

Fig. 4.2. Smallholder farmer in Cambodia ploughing a field. Credit: Dorieke Goodijk.

Intensive land use and high input: large-scale animal production

In most developing countries, the number of large-scale animal production units is increasing in response to surging demand for meat, milk, eggs and other products of animal origin (Fig. 4.4). This is also known

Fig. 4.3. Pastoralist market in Afar, Ethiopia. Credit: Ethiopia Pastoralist Forum.

Fig. 4.4. Intensive pig production in the Philippines, which is increasingly common in the developing world. Credit: Dorieke Goodijk.

as 'the Livestock Revolution' and is based on rapid urbanization and increased purchasing power in many societies. The livestock sector is rapidly moving towards intensive and specialized systems, in which the production environment is controlled and aimed at maximum productivity per animal. Intensification is accompanied by scaling up of production. This is combined with globalized trade of animals and livestock products, like semen, for example.

Specialized pig, poultry and dairy farms around urban areas are mostly large-scale units based on high input of finances and externally derived resources. In these units, only one animal species of a highly specialized exotic breed is raised. Mechanization reduces the labour required with a major role for men. Market targets can be national and international. The single outputs of the system – such as meat, milk, wool or eggs – have the potential to feed vast numbers of people. There is a high use of commercial chemical products for animal health care, fertilization and crop protection. There is also a great risk of environmental damage if control of inputs and outputs gets lax. Compared with the low-input systems, this livestock-keeping system receives a lot of attention from policy makers, research, extension services and education.

Extensive land use and high input: ranching and feedlots

Ranches are generally specialized large-scale units to produce meat from cattle, sheep or goats. Ranches rear animals extensively in large pastures and depend on roughage for a significant portion of the nutrition throughout the growth periods. Feedlots are designed to finish the ranch livestock to a desired market weight and meat quality. This is done either in dry lots with grain-based rations or in more extensive, grass-fed growing operations (Fig. 4.5).

Dry lot or grass-fed meat production units normally operate with high inputs in terms of finances and outside resources. There is little diversity in livestock breeds on ranches. Animals are bred for special characteristics, such as size or age at finished weight, meat quality, good mothering ability, hot weather and insect tolerance, and other attributes that fit the locale or market channel. Livestock production is often accompanied by intense farm crop production and high levels of mechanization. As in other high-input and specialized systems, the focus is on maximizing individual animal productivity. The ranching system attracts much attention from research, extension and education, as well as from policy makers.

The ranching system for beef is responsible for a large portion of the forest destruction throughout the world. Sometimes, forests are cleared for crop production and later on cattle grazing. This can lead to serious environmental damage when not carefully managed. Markets are aimed nationally and internationally.

Fig. 4.5. Zebu cattle in a ranch for specialized meat production in Cuba. Credit: Dorieke Goodijk.

Livestock-keeping Systems in Developed Countries

The four livestock-keeping systems shown in Fig. 4.1 can be identified around the world, though with different characteristics in each region. In northern Europe, the following livestock-keeping systems can be found (with an emphasis on The Netherlands).

Intensive land use and high input: large-scale animal production

Large-scale and conventional (non-organic) high-input farming is most frequent in The Netherlands, and provides most of the milk, meat and eggs. Agriculture in The Netherlands only 50 years ago was quite similar to agriculture in developing countries: family farms with a diversity of activities, combining low-input crop production with various species of livestock and numerous other activities. Since the 1960s, policies are aimed at highest yields through specialization, mechanization and scale enlargement. Production levels were boosted (Fig. 4.6). This also resulted in large-scale export of dairy products (Fig. 4.6).

After several years, however, the negative side-effects also became clear. The high-input farming system led not only to environmental pollution but also to animal disease, and declined farm income. The number of dairy

Fig. 4.6. The high-input dairy system in The Netherlands: this cow has produced over 100,000 l of milk in her life. Experiences with such high-input dairy systems have shown that intensification and scale enlargement have both positive and negative effects.

farmers was reduced by nearly 90% since 1960. The large-scale animal farming sector is also increasingly criticized for reasons of animal well-being, climate change effects, environmental pollution and excessive use of antibiotics, resulting in multi-resistant microbe strains. This has resulted in numerous new (and often farmer-based) initiatives (see also Box 2.10 in Chapter 2).

Besides conventional large-scale farming, a small but growing group of farmers in developed countries are engaged in organic forms of production and management. Some meat, milk and egg producers manage their stock and feeding systems to serve the

Fig. 4.7. Ranching in Denmark with specialized beef cattle.

certified and organic markets. Other types of certifications abound in relation to feeding and management, such as all grass-fed, all natural, humane certifications, free-range, quality assurance, cage-free and specific breed certifications. This is growing in popularity amongst consumers (see also Case Study 6, Chapter 12).

Extensive land use and high input: ranching and feedlots

Meat production is another specialized and high-input sector. Whereas the production of broilers, pigs and calves should be categorized under the large-scale and high-input system, specialized beef production can be classified under more extensive and high input. More extensive beef production (Fig. 4.7) with grazing during the earlier growing phase is followed by a more confined finishing phase.

Improved breed beef cattle are fast growing and produce large quantities of meat per animal. Beef industries have a variety of improved genetics for different climates and conditions that produce good-quality meat that is flavourful and tender. A finishing period on grass or in a confined feedlot with higher grain rations are options for producers.

Intensive land use and low-input: hobby farming

Although in The Netherlands the intensive systems produce most of the food, an increasing number of animals are kept by citizens within an intensive but relatively low-input system: hobby farms. A wide variety of species and breeds is used by individual families or small farms, such as children's farms. These farms are usually located in urban areas and have an educational purpose.

Many families also keep animals for hobby purposes in urban, peri-urban and rural settings. Production is not the main objective; it is more for company, recreation and status. There is also a growing number of citizens organized in breeding associations that aim to maintain special traditional breeds, such as the goat breed in Fig. 4.8.

Fig. 4.8. In Europe, traditional breeds are often maintained by hobby farmers.

Fig. 4.9. Local cattle breeds are kept in low-input systems for the purpose of maintaining natural areas.

Extensive land use and low-input: nature management

Extensive and low-input livestock keeping is a growing phenomenon within The Netherlands, with the special purpose of nature management. Over the past few decades, the traditional sheep and goat pastoralists have vanished, but more recently, a modern form of pastoralism is being developed. Sheep and cattle are used in natural areas to keep the vegetation low, and to re-create a natural appearance. The animals are supposed to survive by themselves or receive limited extra feeding during the winter. In the process, the profession of sheep herding is being revalued: he (or she!) is being paid for nature management. Modern-style sheep-herding also attracts urban people with burnout or other modern lifestyle-related problems (Fig. 4.9).

Reference

Hooft, K. van't, Millar D., Geerlings, E. and Django S. (2008) *Endogenous Livestock Development in Cameroon – Exploring the Potential of Local Initiatives for Livestock Development*. Agromisa Foundation, Wageningen, The Netherlands.

5

Livestock-keeping Systems and Poverty

Learning Objectives: Understanding
- The livestock-keeping systems relevant to the poor
- The role of livestock as part of integrated agricultural systems
- Family strategy of risk minimization and diversification
- The link between livestock and most vulnerable groups
- The influence of livestock on environment and climate change
- Risk and advantages of low-input livestock keeping for human health
- The potential of smallholder mixed farming and pastoralism for Millennium Development Goals (MDGs) and biodiversity conservation

For hundreds of millions of poor households, livestock remain a key asset, often meeting multiple needs and enabling livelihoods to be built in some of the world's harshest environments. Livestock make a vital contribution to food and livelihood security, and to meeting the United Nations Millennium Development Goals. It will be of increasing significance in the coming decades. Traditional livestock keepers – often in poor and marginal environments – have been the stewards of much of the animal genetic diversity and integrated production systems. We should not ignore their role or neglect their needs.

Jacques Diouf, FAO Director General (FAO, 2007a)

Livestock-keeping Systems Relevant to the Poor

According to the FAO, 640 million smallholders and 190 million pastoralists are raising livestock. They make up 70% of the world's poor (FAO, 2009). It is important jointly to analyse the types of animal keeping used within a family and community, before focusing on ways to support the animal husbandry system. Most livestock-dependent resource-poor families can be found in the two major systems: in the low-input smallholder farming (Fig. 5.1) and pastoralist systems.

The type of animal keeping system used within smallholder farming is primarily related to the characteristics of the family and the conditions of their surroundings. Most families combine various crops with a wide array of livestock species. Most of these species are managed under a low-input diversified livestock-keeping system.

Whenever an opportunity presents itself, the family (or one family member) may decide to specialize the keeping

Fig. 5.1. This member of a rural family in Bolivia depends on a keeping a variety of livestock species in combination with growing crops and a number of other income generating activities.

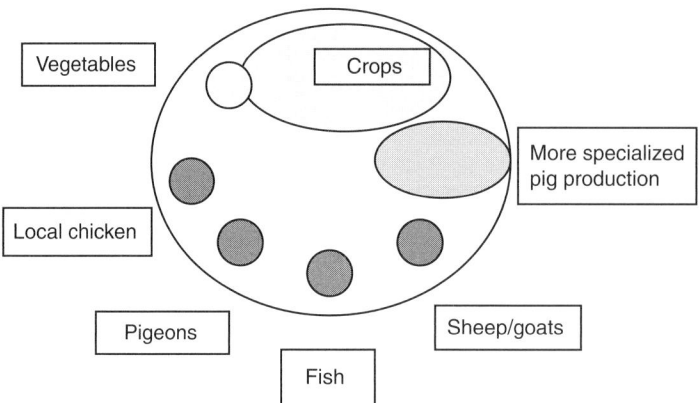

Fig. 5.2. Many livestock-keeping families combine low-input keeping of various species with more specialized keeping of one selected species, in this case pigs.

system of usually one selected species. This species is then managed under the more specialized livestock-keeping system. This will have a lower input level than the large-scale commercialized farms or ranches that are typically managed by large-scale landowners and investors, but in terms of productivity focus, these more specialized systems have adopted some characteristics of the high-input commercial farming systems. The rest of the animal species remain under the low-input conditions (Fig. 5.2).

Therefore, within the smallholder system two kinds of livestock-keeping strategies can be identified: the *diversified and low-input* livestock keeping and the *more specialized* keeping of one selected species (see also Chapter 6, Fig. 6.1).

General Household Strategies

Many families in developing countries live in risk-prone and climatically unpredictable environments. The most effective response in these circumstances is based on productive diversification and risk minimization. Agricultural production is critical in rural strategies because it structures the relations of production around the land. Nevertheless, there are many other sources of income that can contribute to the diversification of the family economy. It is estimated, for example, that 90% of the rural families in central and southern Bolivia obtain more than 50% of their income from non-agricultural activities, including migration.

Life strategies of rural families with scarce resources are thus based on three fundamental aspects:

1. *Crop, livestock and forest production*, generally on a low-input basis mainly for family consumption and some sales at the local market. One aspect within this system, such as a crop or an animal species, may be selected for more specialized production aimed at the market and direct monetary income.
2. *Non-agricultural activities* within the community that generate monetary income. For example, home production of crafts, or agricultural and livestock by-products, buying and butchering animals, transport and trading of agricultural products, opening a store, a bar or a small restaurant.
3. *Migration* and other activities that generate monetary income outside the community. There are many types of migration: permanent, seasonal, sporadic and round trip. They may include only one family member, various members of the same family or the entire family.

The relationship between these three strategies varies greatly among families. In addition, it may vary within the same family depending on the season of the year, age of the family and external circumstances. Migration has become an increasingly important strategy. This has numerous social effects.

Risk Minimization and Diversification Strategies

In the process of facing the challenges, families continue to build on their age-old risk aversion and diversification strategies within their agricultural system. These family and community level strategies continue strong today in many parts of the world. This is expressed in the high diversity of livestock-keeping systems used.

On the one hand, this is expressed in terms of species used: unlike large-scale animal productions systems that are largely limited to three species – chickens, pigs and cattle – families can raise up to ten animal species. On the other hand, there is a large variety between families, regions and cultures in the ways each of these species are managed.

Diversification and risk management play an important role in food security and social relationships within communities, and are often based on the principles of solidarity and equitable redistribution of resources. For example, in Ethiopia, pastoralists have a conscious herd management strategy to reduce risks during lean seasons (Gebru Tegegn, 2009). In the Andes, barter (the exchange of goods) and reciprocity systems continue to be part of the way rural families and communities organize themselves (Box 5.1).

Women and Livestock Development

Recognition of the links between livestock production and hunger, gender inequality and vulnerability to debilitating diseases has helped turn the spotlight on women and livestock in development. Above all, with a view to alleviating poverty, special attention needs to be given to women (Fig. 5.3). In many cases, women are the ones primarily responsible for the livestock. There is an intimate relationship between the situation of farming women and animal husbandry that they practice (Letty and Waters-Bayer, 2008).

Limitations

- Women have many responsibilities, especially under circumstances of

---

Box 5.1. Reciprocity and regional exchange: the Seven Day Fair in Bolivia.

In the valleys of Bolivia, it is still very common to find barter and reciprocity, especially amongst the indigenous population. Barter is common, for example, between families that live in high-altitude zones and the valleys: exchanging cow manure from valley farms for potatoes from high altitudes, for example. This also takes place in traditional fairs, in which practices such as barter and reciprocity are mixed with a money-based market.

The Seven Fridays Fair occurs seven Fridays after Easter in the town of Sipe Sipe, located close to the city of Cochabamba in Bolivia. This period coincides with the root plant harvest on the high plains, and the maize harvest in the valleys. These products are exchanged via barter between families from the valleys and from the high plain zones. The fair is very important for the people in both regions as it contributes to food security. Moreover, it reinforces social relations between family, godparents and friends from different communities. During this fair, various rituals of thanks to nature and to the Mother Earth (Pachamama) are performed, both within and outside of the Catholic Church (Compas, 2007).

Fig. 5.3. With a view to alleviating poverty, special attention needs to be given to women. In many cases, women are the ones primarily responsible for the livestock. There is an intimate relationship between the situation of farming women and animal husbandry that they practice. Credit: Ellen Geerlings.

out-migration of family members, and dealing with the effects of HIV/AIDS.

- Women receive little or no formal education.
- The possibilities of leaving home are limited.
- The amount of land available to them is generally limited by culture or by law.
- They have little or no capital to invest.
- The possibilities of each woman depend heavily on the age of her children; a woman has fewer possibilities with small children than she does without

children or with older children that can help her.

Strategies

- Women seek diversification of the activities in order to minimize risks; for this reason, raising animals plays an important and constant role.
- Women are strong movers of family and community organization.
- Family animal husbandry headed by women is generally an activity of minimal investment with few locally available products.

- Raising animals offers a good possibility of generating monetary income for women, which is usually used for all the family's benefit.
- Working with animals is considered more pleasant than other activities that can yield similar income.
- Animals are part of the indigenous culinary and medicinal culture, intimately tied to the responsibilities of women.
- Many women prefer to have animals around the house as part of their daily environment and for their affective relationship with them.
- Teaching from mothers to daughters is done orally and through the participation in daily activities.

This combination of elements explains why diversified animal husbandry, led by farm women without regard to their social origin, has existed for thousands of years and will continue to exist in spite of its apparent 'irrationality' from a technical–economic viewpoint. Rather it deals with a very profitable activity in socio-economic terms as a farm woman explains:

'I always give food to my pig. I invest a lot of effort in this and I do not care if it is economically profitable. The most important thing is that I can sell it when my children need books for their classes'

(van't Hooft, 2004)

Children also have an important role in low-input livestock systems, especially in communities where the animals are taken out to graze. When these children enrol in school, manual labour becomes scarce and has a negative impact on the workload of women. At the same time, shepherding work inhibits the children in their formal education. This affects girls especially.

Smallholder Livestock Keeping, Environment and Climate

The role and potential of livestock in influencing environment and climate change is subject to much debate. There can be major ecological problems related to raising animals, such as excess nutrients and waste, the use of large quantities of grains at the expense of crops for human consumption and the associated emission of greenhouse gases such as carbon dioxide (CO_2), methane (CH_4) and nitrous oxide (N_2O). Livestock had not been mentioned extensively in the climate change debate until 2006, when the FAO reported that livestock keeping produces 18% of all greenhouse gases (Steinfeld et al., 2006).

In the climate change discussion about livestock, the focus is often on ruminants, and on specialized ways to maximize individual animal productivity. High-input solutions promoted to reduce the greenhouse gas emissions by ruminants include feeding ruminants with more maize and concentrates, change from keeping ruminants to pigs or poultry, change local breeds for more productive and specialized ones, build totally enclosed stables, reduce methane production in the rumen. To achieve this, it is often argued that intensification and large-scale livestock production is necessary.

However, there is a need to take into account the total environmental problems related to livestock production, including water scarcity, land degradation, loss of soil fertility and loss of biodiversity. A differentiation should be made between existing livestock-keeping systems, in terms of climate change and environmental impact, especially between high-input industrial systems and low-input systems (van't Hooft, 2009).

Box 5.2. The Women in Livestock Development (WiLD) Initiative.

The WiLD initiative empowers women by creating opportunities to own livestock. Heifer International provides women with cows, goats, buffalo or poultry, resources for livestock production, values-based literacy and gender-equity training to strengthen women's positions in the community. Given the opportunity, women generate and handle income for the benefit of their families and get involved in planning and analysing the outcomes of the women-focused activities. (www.heifer.org/wild)

Experiences have shown that there is also another option: the optimization of livestock-related systems as a whole, which reduce the negative effects of livestock keeping, while maintaining levels of total farm production (Sere, 2009). Small-scale mixed-production systems that combine crops with raising animals offer good possibilities for balancing food production with sustainable replacement of nutrients, and thus a better chance of maintaining the fertility of the land. The social and environmental value of small-scale livestock production systems can far outweigh the negative consequences. Similarly, there is increasing recognition that pastoralist systems may well be the most appropriate form of utilizing grazing resources under arid conditions.

Positive effects of low-input family animal husbandry on environment and climate

- *Increased soil fertility and organic matter*: The integration of crops and livestock in family production helps to preserve the fertility of the land through the recycling of nutrients. The use of manure on plants preserves the structure of the soil and its drainage capacity. The need for an integral productive system is especially necessary in the humid tropics because of its fragile ecosystem and acidic soil. Increased soil fertility and organic matter also increase the absorption of CO_2 because of plant growth, which can be a significant factor in terms of mitigating climate change effects of livestock (Alford and Penney, 2006).

- *Animals generate energy*: Animals produce energy through transport, draft power and the conversion of biomass into foods. This reduces the demand for fossil fuels (Fig. 5.4). Manure is used directly for fertilizing the fields, for cooking, heating and generating electricity. In addition, manure from some species can be used to feed other species; raising fish, for example, can be done with cow manure, pig and duck manure.

- *An alternative source of income*: As an alternative to monetary income for the family, animals lessen the need to carry out other activities that cause greater ecological harm, such as cutting trees to sell charcoal or slash and burn to create farmland in tropical zones within virgin forests.

- *Animal biodiversity*: Rearing livestock that are native to a region makes sense from the standpoint of acclimatization. Long-term resident breeds and strains are adapted to weather conditions, insect pests and natural feed variations. When we pay attention to raising these animals for production, work and by-products, we take advantage of these natural characteristics. Most regions of the world have a variety of different breeds of

Fig. 5.4. Animals produce energy through packing, draft power and the conversion of biomass into foods. This reduces the demand for fossil fuels. Manure is used directly for fertilizing the fields, for cooking, heating and generating electricity. Credit: Jeet Lal Shrestha.

animals that provide this broad diversity. Finding ways to utilize these variations makes biological sense. Diversified breeds and types of animals in the production system also allow families to utilize more fully the variety of crops that grow in marginal lands. Unfortunately, many local breeds of animals – both domestic and wild – are in very low number and some are in severe risk of being lost completely. Therefore, selection of local breeds for family-level livestock production can preserve this genetic diversity and enhance local food security (FAO, 2009).

In smallholder and pastoralist systems, local breeds are commonly used (Fig. 5.5). Even though their production of traditional products (meat, milk and eggs) is relatively low, their efficiency is high, as they produce on the basis of low-quality foods and low costs. These breeds are also adapted to the existing eco-cultural factors of the zone.

- *Use of by-products*: Livestock can take advantage of many animal and plant by-products in their feeds, such as low cost and waste products from slaughterhouses and restaurants. Even chicken manure from the poultry industry and remains from the fish and fruit industries have beneficial nutritional uses. Thus, environmental contamination from these low-value remains is avoided.

Fig. 5.5. Cattle at Fulani cattle market, Cameroon. Photo credit: Ellen Geerlings.

Negative effects of low-input family animal husbandry on environment and climate

- *Overgrazing*: A main source of environmental degradation and greenhouse gas production related to livestock is related to poor land use, like soil degradation, overgrazing and deforestation. The degradation of natural grazing fields is related to poverty, overpopulation of the zone and ecological changes. As a life strategy, many rural families in these zones are forced to satisfy their immediate needs through unsustainable practices, such as overgrazing and deforestation for the sale of wood for cooking and charcoal. In zones where there are many small properties, there is the danger of degradation by a growing population. When the parcels will no longer support livestock, the fertility of the land decreases rapidly and worsens the situation of degradation. Over time, this becomes a vicious cycle of poverty and degradation. Livestock have a determined influence even though they may not have been the initial factor.

- *Slash and burn*: Increasing numbers of people in tropical forest zones are also responsible for considerable environmental damage, especially if they have migrated from other ecological zones. This is where we see large quantities of virgin forests taken down each year by smallholder farmers in search of land for producing food and cash crops.

- *Goats*: In some dry and deforested zones, many families keep large herds of goats. These are seen by outside agents as a 'curse' because of their foraging on the small amount of shrub vegetation in the existing grazing lands of these zones (Fig. 5.6). Yet at the same time, families consider these animals to be a blessing because they depend on them for income in their difficult life conditions. Goats (temporarily) prevent a definite out-migration of these families.

Fig. 5.6. Goats are often managed in a different way from sheep and cattle. They prefer to browse trees and bushes more than grasses.

Smallholder Livestock Keeping and Human Health

In developing countries, the majority of families' livestock keeping takes place in low-input conditions. In the cultures and strategies of these families and communities, there is an intimate relation between animals and humans, at both physical and emotional level. This way of living together with animals has a lot of consequences.

In general, the theme of the relation between animals and humans does not get much attention in agricultural curricula. Meanwhile, the diseases that pass between animals to humans (zoonoses) have serious consequences for public health, and in the majority of cases, they have not been controlled in an effective way. In communities and families, the knowledge on diseases and parasites that pass back and forth between animals and humans is also very limited. Because of this lack of information, there are many beliefs about these diseases that are not always based on reality.

Advantages of low-input family animal husbandry on human health

There are numerous advantages to keeping animals, which explains how we can find this activity in most parts of the world; among rich and poor people and in rural and urban areas. Animals are related to the well-being of families in the following ways:

- *Optimization of agricultural production and farm efficiency*: Within the strategies of the rural families, agriculture production is intimately related to livestock, so that the animals take over the draft power and transportation needs of the family. They save on physical efforts of the family. In many cases, manure is the most important way to fertilize agricultural fields; moreover, the animals use the agricultural by-products.
- *Income possibilities*: Families sell animals or their products in moments of necessity or emergency. With this money, they can pay for a visit to the doctor; buy school materials or other products (bread, clothes, medicines, cover funeral costs). These elements are directly related to the well-being of the family.
- *Animals provide food for the family*: The production of food products (milk, meat, eggs, honey and fish) is an important reason to keep animals. Independent of direct nutritional benefits, foodstuffs increase physical health of the family. The diseases, parasites and other elements that cause physical stress affect less if the physical body is well nourished.
- *Use of household organic waste*: Several animal species, such as chickens, pigs, guinea

pigs, rabbits, dogs and cats eat leftover food and help to control rats, mice and insects. In this way, they contribute to the physical cleanliness of the household environment.

- *Physical protection*: Animals such as dogs, geese and guinea fowl play a role in protection of the family, both during the day and in the night, so the family can sleep without worry. The wool and fibre of different species also serves as protection against the cold.
- *Emotional and spiritual comfort*: Animals form a part of many cultural festivities, ceremonies, rites and social events: This is important for cultural identity, which influences emotional stability, both at the individual and community level. Moreover, animals provide companionship and are part of everyday life and contentment.

Disadvantages of low-input family animal husbandry on human health

There are also disadvantages of keeping animals. It is necessary to understand them in order to decrease the risk.

- *Social problems*: Problems can arise with neighbours related to the theft of animals or over animals that are noisy or left to run loose and that disrupt crop production. This can create tension and disagreements between families.
- *Hygiene*: Where animals are kept and where milk, meat and animal feeds are stored, there is more humidity and odours which attracts flies, cockroaches, rats and other vermin that feed on scraps. To this, we must add that when animals are loose and in the house, they may defecate in the house. Animal waste can attract additional pests.
- *Medicinal residues in milk and meat*: In family livestock keeping, medicinal compounds such as antibiotics and hormones are often used. Product labels specify proper use and withholding times from consumption of milk, meat

or eggs. In order to avoid residues in human foods, these label instructions must be followed carefully. Often this is not the case and as good control is lacking, milk and meat are often contaminated with toxic residues. A side-effect is that microbes and other disease-causing agents can develop resistance against the antibiotics. This results in so-called multi-resistant strains of microbes in both animals and humans. In cases of disease with these strains, no antibiotic treatment will be effective.

- *Zoonoses*: Zoonoses are diseases or parasites that pass between animals and people. They represent a problem between animal husbandry and human health, especially in places where animals and people live physically close together.

Low-input Livestock Systems and the Millennium Development Goals

As shown in the various parts of this chapter, low-input livestock-keeping systems have a role to play in most elements of livestock keepers' lives. They are thus also related to most of the eight MDGs, but mostly in the following (FAO, 2007b; Heifer International, 2010):

MDG 1: reducing hunger and poverty

It is increasingly recognized that livestock in integrated systems can contribute positively to reducing hunger and poverty, thus addressing MDG 1. Since agriculture is the livelihood source for the majority of the rural resource-poor, investing in this sector will reduce extreme poverty and hunger. Because livestock provide diverse goods – including food, draft power, organic fertilizer, and economic and social security – to smallholder farmers, investment priorities should consider livestock-based agriculture within ecologically sound systems.

MDG 3: Reducing gender inequality

Experience has shown that improvements and low-cost investments in small-scale livestock keeping – a dairy cow, a few goats, a few chickens or guinea pigs – offer opportunities for women not only to increase household income but also to control a larger portion of it, thus reducing gender inequality.

MDG 6: Reducing impact of HIV/AIDS

Improved livestock-keeping practices and production, both for home consumption and the market, diversification in income sources from livestock and a stronger position of women as livestock owners help to reduce their families' vulnerability to the impacts of HIV/AIDS and other diseases, thus contributing to MDG 6.

MDG 7: Ensure environmental sustainability

To catalyse environmental stewardship, secured access and ownership of land and productive assets is critical. Insecurities in tenure and access arrangements can aggravate resource over-exploitation. Special attention needs to be paid to gender-differentiated resource management priorities. Climate change adaptation and mitigation strategies must also focus on the needs and capabilities of the resource-poor.

References and Further Reading

Alford, R. and Penney, S. (2006) Preparing to climate proof. The next challenge for Africa's rural poor. Report Send a Cow. The Foundation Series: 'Passing On' learning no. 1, Send a Cow, Bath, UK.

Compas (2007) Strengthening local markets in Bolivia. In: *Learning Endogenous Development – Building on Bio-Cultural Diversity*. Practical Action, Rugby, UK.

FAO (2007a) *The State of the World's Animal Genetic Resources for Food and Agriculture*. FAO, Rome.

FAO (2007b) *Animal Health and the Millennium Development Goals*. Animal Production and Health Division, FAO, Rome.

FAO (2009) Livestock keepers: guardians of biodiversity. Animal Production and Health Paper no. 176. FAO, Rome.

Gebru Tegegn, G. (2004) Herd accumulation: a pastoral strategy to reduce risk exposure. Research Brief 04-05. GL-CRSP PARIMA, Ethiopia.

Heifer International (2010) Approaching the MDGs – Working with Resource-Poor Farmers – Policy Brief. Heifer International, Little Rock, Arkansas.

Hooft, K. van't (2004) *Gracias a los Animales – Crianza Pecuaria familiar en America Latina, con casos de los Valles y el Altiplano de Bolivia* (*Thanks to the Animals – Family Level Livestock Keeping in Latin America with Case Studies from the Bolivian Valleys and Highlands*). Plural Publishers, La Paz, Bolivia.

Hooft, K. van't (2009) Livestock Friend or Foe, The need to look at different production systems in the debate about livestock & climate change. Available at Endogenous Livestock Development Network, http://www.eldev.net/

Letty, B.A. and Waters-Bayer, A. (2008) Recognising local innovation in livestock-keeping – a path to empowering women. Paper in WCAP. World Conference of Animal Production, Cape Town, 23–28 November 2008.

Seré, C. (2009) No simple solutions to livestock and climate change. International Livestock Research Institute (ILRI) Paper no. 10, November, Kenya.

Steinfeld, H., Gerber, P., Wassenaar, T., Castel, V., Rosales, M. and de Haan, C. (2006) *Livestock's Long Shadow – Environmental Issues and Options*. FAO, Rome.

6

Characteristics of Smallholder Low-input and Diversified Livestock Keeping

Learning Objectives: Understanding
- The characteristics of low-input and diversified livestock keeping
- The challenges and potential of this livestock-keeping strategy

As indicated in Chapter 5, within the small-holder system, two kinds of livestock-keeping strategies can be identified: the *low-input* and *diversified* livestock keeping and the *more specialized* livestock keeping (Fig. 6.1) of one selected species.

In this chapter, we will focus on the first of the two strategies: low-input and diversified livestock keeping. The characteristics, advantages and challenges of more specialized livestock-keeping strategies are presented in Chapter 7, whereas in Chapter 8 the process of transition from a low-input to more specialized strategy is described.

Characteristics of Low-input and Diversified Livestock Keeping

Smallholders and pastoralist systems are the most frequently used farming systems in the world, especially by the rural poor, and most of them are based on low-input and diversified strategies. Low-input and diversified livestock keeping has existed for thousands of years and still holds strong

today. It forms the basis of a relatively effective and sustainable production system under circumstances of isolation, insecurity and change. The following are the major characteristics of this form of animal husbandry (van't Hooft, 2004).

Manual labour often from women and children

Many women are alone with their younger children during a good part of the year with the responsibility to carry out the activities of animal husbandry (Fig. 6.2), crops and chores at home. This can be because of seasonal work or migration of their spouses or sons, or loss of family members because of disease.

Logic of low investment and low productivity of food products

This system of animal husbandry is characterized by little investment of money and

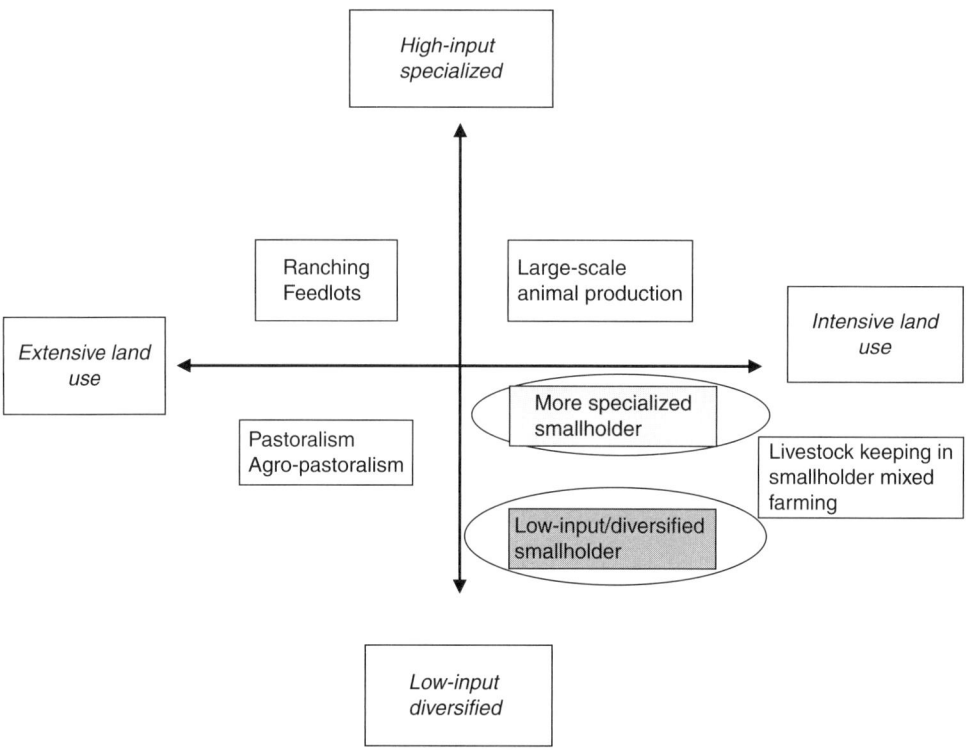

Fig. 6.1. Within the smallholder system, two kinds of keeping systems can be identified: the diversified and low-input system, and the more specialized keeping of one selected species.

Fig. 6.2. Many children cannot attend school because they have to take care of the animals or do other chores. This is especially the case when the animals have to be herded.

manual labour, resulting in incomplete management from the technical point of view. As a result, the per animal production with respect to food products such as meat, milk and eggs – and *expressed in product per*

animal per day or year – is lower than more specialized systems. At the same time, mortality from malnutrition, lack of protection, parasites and infectious diseases is usually much higher. The concept of 'productivity' in this husbandry system is based on the numbers of live animals rather than the level of per animal production per day or year. That is why improvements in this system need to be focused on reducing mortality (Chapter 9).

High potential for improved poverty alleviation and food security

It is important to appreciate the potential of the low-input and diversified livestock-keeping systems in terms of poverty alleviation and food security. The low production from animals is achieved with minimal costs, which translates into relatively higher net gains and significant returns (Fig. 6.3).

Fig. 6.3. Smallholder low-input livestock systems have their own ways and strategies of marketing. Chicken and ducks are raised with minimal costs and sold very easily for a good price in Cambodia.

Moreover, the lack of attention to this system in the past leaves much room for improvement. If optimized effectively (Chapter 9), low-input and diversified systems can be extremely efficient systems, especially in marginal conditions (ANTHRA, 2008).

Raising different species of animals at the same time

In order to reduce the risks of husbandry, different species are often utilized together, each one with specific functions. When one species is affected by disease, for example chicken because of Newcastle disease, other species – such as ducks – will survive (Fig. 6.3). Raising various species together is therefore a conscious risk-spreading strategy. It is also important to emphasize that value of each species of

animal is relative, depending on many factors such as climate, availability of food, cultural and personal factors, the quantity of animals and disease.

Wide variety of animal husbandry forms

Within low-input and diversified animal keeping, there are many different ways of raising each species of animal, depending on the ecological zones, market opportunities and labour available. For example, there is not just a single way of raising pigs; rather there is much variation depending on the characteristics of each family, and their resources and experience. Below are some examples from livestock keepers in the Bolivian Valleys (van't Hooft, 2004).

- Many families raise one or two pigs to fatten and sell when they need cash, for example for schoolbooks for their children. These 'piggy bank pigs' are fed with leftovers from the harvest and the kitchen. Other families raise a larger number of pigs because they operate a local chicha (local maize drink) brewery, which is generally completed with a purchased supplement, such as rice bran; for example, families that make chicha at home, or with a restaurant.
- In the case of sheep, many families in the valleys of Bolivia have some sheep leashed out and take feed to them or they leash them around agricultural fields. In high-altitude zones, families generally graze their medium or large flocks every day.
- Similarly, there is much variation in diversified husbandry of poultry. The majority of farm families keep chickens around the house, or in other instances, ducks, turkeys or geese for their own consumption. Other families raise a certain quantity of chickens or ducks with leftovers and purchased grains to sell part of their production in the market. It is also common to raise semi-wild doves in the valleys of Bolivia to take advantage

of their meat, which has culinary and medicinal values. Intensive husbandry of fighting cocks is usually an activity of men.

• In the case of bees, some families raid wild hives, extracting honey once a year – destroying the hives in the process. Other families raise the bees in rustic hives to take advantage of the honey several times a year. Others purchase improved hives or have adapted their rustic hives with elements of improved hives in order to sell the honey and by-products such as pollen, wax and propolis.

Predominant use of local breeds

In low-input and diversified smallholder and pastoralist systems, local breeds are most commonly used, because they are best adapted to local circumstances. Even though their production of traditional products (meat, milk and eggs) is relatively low, their efficiency is high, as they produce on the basis of low-quality foods and low costs (Fig. 6.4). These breeds are also adapted to the existing cultural factors of the zone, standing at the basis of local dishes, rituals and cultural expressions (FAO, 2009).

Use of non-traditional by-products

By-products from the animals such as wool (Fig. 6.5), manure, blood, hide, bones, fat, horns, intestines, bladders, fetuses and feathers are utilized for a multitude of uses in the home, in the kitchen, for marketable crafts, and for cultural or medicinal purposes. In some cases, animals are raised especially for one of these specific uses.

Limited relationship with the monetary market and technical services

Low-input and diversified husbandry is based on locally available products and is directed toward personal family consumption and the sale of some surpluses. For this reason, the relationship between this type of husbandry and the monetary market is limited. In the case of necessity, sale or barter, the latter is done via vendors that come to the house or by taking the animals to the local fair. The consumer sometimes prefers products from local breed animals rather than products originating from specialized husbandry. The use of technical services in diversified husbandry is generally limited to large animals such as cows, in situations in which the animal is at risk of

Fig. 6.4. Local Chiapas sheep thrive relatively well under local conditions in southern Mexico. Projects to replace these sheep with exotic breeds have resulted in failure. Credit: Paul Perezgrovas Garcia.

losing its life. Measures to prevent diseases or to increase per animal production are usually applied to animals that generate monetary investments, such as cattle or pigs, or fighting cocks in certain cultures (Mathias *et al.*, 2010).

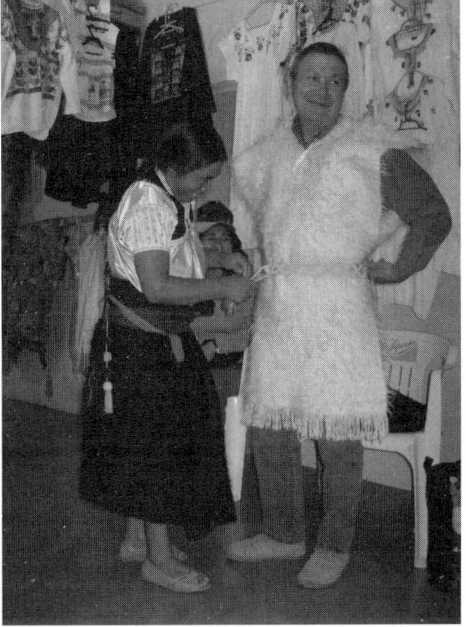

Fig. 6.5. Shepherdess in Chiapas, Mexico, selling a garment made of the special wool of her Chiapas sheep.

Strong interdependence between crops and animal husbandry

Diversified animal husbandry is combined with crops that are fertilized with manure, animals used for draft and transportation (Fig. 6.6). In addition, animals convert agricultural by-products into high-value food products. Leguminous crops for the animals, such as lucerne, are important elements because the plants stimulate the fertility of agricultural lands.

Husbandry based on farmer and ethno-veterinary knowledge

Diversified animal husbandry is based on the experiences and knowledge (Fig. 6.7) generated over generations in families where the traditions have been practised. The use of western science, such as vaccinations and commercial veterinary medicines, is limited and these practices are utilized more with species that generate monetary income (FRLHT and Tanuvas University, 2010).

High flexibility through the purchase and sale of animals

There is high flexibility in this type of husbandry because diversification allows different

Fig. 6.6. In smallholder agriculture, crops and livestock are effectively combined. Leguminous fodder and maize are intercropped; animal manure is used for fertilizing the maize field.

Fig. 6.7. Exchanging knowledge about the use of ethno-veterinary practices is one of the ways to effectively support low-input livestock keeping without increasing the costs. Here local healers of the Cameroon Ethnovet Council of Experts show their experience on the use of medicinal plants. Credit: M.N.B.Nair.

strategies according to the situation that is presented. For example, in the case of needing medicine for a sick child, a chicken can be sold; and when a lot of money is required for a burial, a cow can be sold. When there is an infectious disease in the chickens, it is common to sell all of them in order to prevent greater losses. When the disease has passed, chickens can be bought again. Or ducks can be raised, which are more resistant to infectious diseases.

Reduced flexibility in the case of big changes

When families experience radical changes in their surroundings, low-input and diversified animal husbandry does not adapt easily. For example, in Bolivia, families that have migrated from the high plain to the tropical zones to plant coca leaf generally keep their chickens without any type of protection against the rains, as is the custom in their places of origin. Families originating from the tropical zone protect their chickens with simple, effective constructions. Sometimes, recently arrived families learn to adapt their management strategies to their new surroundings, but often this is not the case because of limited exchange between the two population groups (van't Hooft, 2004).

Limited formal attention

Unfortunately, it is still common to find a lack of information combined with a level of disdain from research and formal education institutions towards low-input and diversified agricultural systems. This leads to inadequate attention to these systems (Mathias, 2010). Moreover, when government services are privatized, this results in further neglect of these agricultural systems. In addition, there are few medications and products adapted to this type of husbandry, and if available, their quality is often questionable. Commercial products, such as vaccines, parasite medicines, vitamins and minerals often come in large quantities that are uneconomical and difficult to use for smallholder livestock keepers.

Vicious cycle of poverty and small land holdings

Under conditions of growing pressure on the land and low soil fertility, diversified agricultural/livestock production can enter into a vicious cycle. When the number of animals necessary for crop fertilization cannot be maintained because of lack of land, the fertility of the land decreases rapidly to the point that agricultural production goes below subsistence levels. Certain species are culturally identified with poverty and indigenous

---

Box 6.1. The risk of rabies (van't Hooft, 2004).

Rabies is officially recognized as a neglected zoonosis by the World Health Organization (WHO, 2005). Rabies is caused by a deadly virus that can be carried by any warm-blooded animal. In developed countries, it is found most commonly in wild animals, such as coyotes, foxes, raccoons, bats and skunks. In less developed areas where vaccination does not cover all dogs, these canine species are most often the biggest threat to humans. In many rural and urban settings, there are large numbers of dogs that mainly stay on the streets. Rabies can also infect livestock, in both low-input and more specialized systems. Cattle are thus another source of infection for humans. Infection is also possible through the bite of a vampire bat.

In such regions, it is not uncommon to have cases of rabies in humans each year. Once a person begins showing symptoms of rabies, the disease is practically always fatal. Thus, a regular vaccination programme for dogs is a very important community public health practice. In areas of frequent problems with rabies, the same vaccine is also administered to cows and other farm animals.

---

populations; in Bolivia, for example, these are goats, llamas and guinea pigs. This may hamper the efforts to work with these species.

Includes urban livestock keeping

Residents of urban areas, especially those in poorer neighbourhoods, often keep animals to support their food security and income (FAO, 2001). In spite of many regulations to limit this kind of animal keeping, it is common to see animals grazing at the roadside or at the backyard in most smaller and major cities in developing countries. Most of these animals are kept in a low-input and diversified system.

Risks to human health

In diversified husbandry, because of the close contact between animals and humans, there is a special danger of infection from disease and zoonotic parasites (WHO, 2005). This phenomenon has a direct relationship with the conditions of poverty and lack of hygiene, as in the case of some serious and common zoonoses, such as cysticercosis, rabies, tuberculosis and (in South America) Chagas disease (Box 6.1). In addition, raising animals close to or inside the house attracts numerous undesirable insects and rodents, such as flies, mosquitoes and rats.

References and Further Reading

ANTHRA (2008) Unpacking the poor productivity myth. Women resurrecting poultry biodiversity and livelihoods in Andhra Pradesh, India. Good Practice Brief SAGP25, South ASIA Pro-Poor Livestock Policy Programme, FAO/NDDB, Rome.

FAO (2001) Livestock keeping in urban areas. A review of traditional technologies based on literature and field experience. Animal Production and Health Papers, no. 151. FAO, Rome.

FAO (2009) Livestock keepers: guardians of biodiversity. Animal Production and Health Paper no. 176. FAO, Rome.

FRLHT and Tanuvas University (2010) *Proceedings of International Conference on Ethnoveterinary Practices. Mainstreaming Traditional Wisdom on Livestock Keeping and Herbal Medicine for Sustainable Rural Livelihood across Continents*, Thanjavur, Tamil Nadu, 4–6 January.

Hooft, K. van't (2004) *Gracias a los Animales Crianza Pecuaria familiar en America Latina, con casos de los Valles y el Altiplano de Bolivia (Thanks to the Animals – Family Level Livestock Keeping in Latin America with Case Studies from the Bolivian Valleys and Highlands)*. Plural Publishers, La Paz, Bolivia.

Mathias, E. (2010) Mainstreaming ethnoveterinary medicine in veterinary education and research. In: *Proceedings of the International Conference on Ethnoveterinary Practices*, Thanjavur, Tamil Nadu, 4–6 January.

Mathias, E., Mundy, P. and Köhler-Rollefson, I. (2010) Marketing products from local livestock breeds: an analysis of eight cases. *Animal Genetic Resources* 47, 59–71.

WHO (2005) The control of neglected zoonotic diseases – a route to poverty alleviation. Report of a Joint WHO/DFID-AHP Meeting. WHO, Geneva.

7

Characteristics of Smallholder More Specialized Livestock Keeping

---

Learning Objectives: Understanding
- The characteristics of more specialized livestock-keeping strategy
- The challenges and potential of this livestock-keeping strategy
- The positive factors and cautions of moving from low-input strategies into more specialized systems

In smallholder more specialized livestock keeping, the livestock-keeping family has selected one animal species on which to focus in terms of labour and inputs, with the expectation of receiving greater monetary return. The focus is on a few animals that have advantageous market characteristics and that require some intensity of capital and labour. As explained in Chapters 5 and 6, this family may have diversified crop and livestock production as well as more specialized productive animals of one selected animal species.

More specialized smallholder livestock keeping is based on the logic of investing in some of the animals to achieve higher per-animal production, with a stronger focus on selling animals and by-products in the market. This also implies the need for increased care and a greater risk of loss when things go wrong. In these more specialized systems, several families often band together to raise the same animal species in an improved way and capture economies of scale in shared husbandry practices, production of feedstuffs and other resources, as well as group marketing for better returns.

The following are the major characteristics of this more specialized livestock-keeping strategy (van't Hooft, 2004).

Labour from the Whole Family

The men of the family usually play a more important role in more specialized livestock keeping (Fig. 7.1) than in low-input diversified livestock keeping. He is usually supported by his wife(s) and children. These men may have a day-labour job in the community that may not meet the minimum survival requirements of the family. Occasionally, women without young children have enough time to dedicate themselves to this type of animal husbandry, or it can be accomplished with paid workers when capital is available from other activities.

©K.E. van't Hooft, T.S. Wollen and D.P. Bhandari 2012. *Sustainable Livestock Management for Poverty Alleviation and Food Security* (K. van't Hooft, T. Wollen and D.P. Bhandari)

The Logic of Investing to Obtain Higher Earnings

Dedicated and sometimes significant amounts of money and time are needed when a single species or breed of livestock is chosen for more specialized keeping. Many times, the family counts on other sources of income to make this possible, for example from marketing livestock, running a small café, providing transportation, construction, food processing or other farm trades. The investment is used to improve animal nutrition along with other forms of sufficient general care. This is expected to reduce mortality and improve production of traditional products, such as traction, milk, meat, fibre/wool, eggs and marketable offspring. The monetary return to the family can be significant to pay debts, provide family needs and improve the home and farm infrastructure. Generally, several years of adaptation are needed in order to fully benefit from the specific conditions and challenges of this new and more specialized form of animal husbandry.

Possibilities for Avoiding Migration

Specialized animal husbandry is one of the possibilities for increasing family income in such a way that migration by one of the family member wage earners becomes less necessary (Box 7.1). In this way, social disintegration of families can

Fig. 7.1. In this more specialized smallholder zero-grazing dairy system in Tanzania, men as well as women are employed and earning an income.

Box 7.1. Avoiding migration through micro-credits and animal production.

Helen and Golly Lagnas with their five children live in the rice paddies on the island of Mindanao in the Philippines and are emerging out of the poverty that grips families of this region. As members of a Heifer International Project community group, they received a micro-credit loan and purchased some pullets and roosters to rear on their small plot.

Since Golly works for day wages in town, Helen and the children manage the growing chicken flock. The family now consumes about 20 eggs per week and eats poultry meat at least twice. In addition, they sell eggs and surplus hens in the market and have passed on pullets to other families in the community. These neighbours are also now eating more protein and receive income from selling eggs.

Micro-credit loans live on through the ability of the family to use money from the sale of produce to scale up their farms. Helen and Golly later purchased some doeling goats and now raise goats for sale. Their next venture into farming will be with the purchase of a young female pig and then a calf. When asked about the change in their livelihood, Helen said that their children are now healthy and the older ones are now going to school. The chickens and goats pay for all of this (Wollen, 2011, personal communication).

be avoided by the out-migration of core wage earners of the rural population. This has impact on the individual family as well as the broader community. At the same time, income from migration is also used to get started. Some families first generate money from migrant or seasonal work abroad, to prepare themselves for more specialized livestock keeping at home.

Greater Uniformity in Husbandry Practices

When moving into specialized livestock keeping, there is more uniformity in the equipment, feeds and husbandry practices used than in the case of low-input diversified animal husbandry. The variations are based on the characteristics and opportunities in the area and the experience of each individual family. In this process, innovative combinations of local and modern practices are found (Fig. 7.2).

Use of Specialized Breeds, or Crosses with Local Breeds

To start off this more specialized system and accomplish 'quick wins' in terms of productivity, specialized or 'exotic' breeds that originate outside the region are often introduced. This does carry certain risks that must be well understood before putting in

the effort, equipment and other needs, and related extra costs. Each individual animal of the exotic breed represents a high value and its husbandry requires special care. If adequate nourishment during the dry season and control of parasites and other diseases is not guaranteed, the fertility problems and mortality among these animals are often far higher than with local breeds. Therefore, the outcome and economic results of this type of husbandry depends greatly on the quality of management (FAO, 2007).

Greater Risks and Less Flexibility

The animal species selected for specialization has to produce sufficient income to cover the greater costs of inputs for the animal as well as the family needs. The chosen animals must maintain a certain level of production to stay ahead of the expenses incurred. The risk involved is greater than in the case of low-input diversified husbandry, because more specialized livestock keeping depends more heavily on external conditions, such as markets, that are not under the direct influence of the family. The risk tends to be especially large in the first few years when facilities and equipment need to be purchased and new sources of feed need to be grown (Fig. 7.3).

In addition, it takes time to acquire the necessary new knowledge and experience related to this type of animal husbandry,

Fig. 7.2. Many families with more specialized livestock keeping effectively combine this with management practices of low-input systems (Vietnam).

Fig. 7.3. The risk involved in more specialized animal keeping is greater than in the case of low-input diversified husbandry, because of the high value of the individual animal as well as the inputs used.

disease protection and marketing. One cannot react quickly to situations of external change, because each animal has a high monetary value and interventions may require costs that are not easily covered. For this reason, families usually have other species as additional sources of income. When unexpected needs arise, these additional animals present a more ready asset that can be turned into cash.

Proximity to Roads, Towns and Markets

More specialized livestock production requires a close relationship between families, the market and technical services in order to be able to procure the necessary inputs for husbandry and for the sale of products. For this reason, family farms and livestock holding facilities are often located close to roads and in the proximity of processors as well as markets.

Less Interdependence between Crops and Raising Animals

As husbandry practices are changed to accommodate more high-yielding animals, the relationship between animal husbandry and crop agriculture production changes. It is possible that more of the feed will need to be purchased from the market since certain types of feedstuffs cannot be grown or provided from the farm. This increases the need for improved animal production to cover the cost of outside inputs and, therefore, the risks on the entire farm. With some types of animals, such as high-yielding breeds of pigs, layer hens or broiler chickens (Fig. 7.4), close to 100% of the feed derives from the market in order to meet complete feed requirements.

Formal Attention

For animal health care providers, such as private or government veterinarians, it is easier to work with specialized husbandry systems than with a diversified low-input livestock keeping system. The logic of more specialized type of husbandry coincides with the content of their training, while the technician can work with men, and in this way avoid the need for learning the local language. In addition, medications and other products whose quantity and quality is adapted to this type of husbandry can be obtained.

Importance of Farmer Associations and Cooperatives

More specialized livestock keeping will often result in a greater abundance of livestock products. A trade union or organization of affiliated families is often promoted by projects to facilitate training, marketing, getting through difficult times and other

Fig. 7.4. Interdependence between crops and livestock is reduced in the case of more specialized pig and chicken production. In the case of producing broilers in Nepal, for example, most feeds are purchased from the markets and not produced on the farm.

Fig. 7.5. Association of Fulani pastoralists in Cameroon. Credit: Ellen Geerlings.

external factors. Being part of an organized group of farmers can reduce risks and can help in market negotiations (Fig. 7.5). When difficulties in the community arise because of the differences between the conditions of affiliated families and those that are not affiliated, the association can help to find win–win solutions.

Environmental and Human Health Impacts

More specialized systems often require more technical and chemical inputs and this requires special awareness of their environmental impacts. Environmental problems associated with the abuse of more specialized (and especially the large scale)

livestock production are too common, though different from the environmental problems resulting from low-input systems. It is often the process of concentration of large number of animals that results in difficulty (FAO, 2009).

- A greater concentration of animals in smaller holding areas results in dust, odours, insects and rodents.
- More specialized animal keeping often requires crop chemicals to control disease and insect pests in special feed crops. These chemicals may run off into adjacent fields or into streams and rivers.
- The use of chemical fertilizers puts pressure on soil nutrients, over time resulting in lower soil fertility.

- More specialized crop and animal production requires more water. When irrigation from deep wells is used, the water can leave excessive minerals on the soil surface that eventually leads to salinization and soil erosion.
- The wastewater from slaughter and processing plants can pollute water streams.
- The indiscriminate and inadequate use of antibiotics for treatment of disease and hormones as growth promoters can lead to residues in milk, eggs and meat. Besides low quality of these products, this can lead to multi-resistant strains of microbes and viruses. This results in serious human and animal health problems.

Zoonotic Diseases

Diseases that are transmitted between animals and humans are called zoonotic diseases and include serious maladies, such as rabies and tuberculosis. Although less frequent than in low-input diversified systems, zoonotic diseases can persist in more specialized systems too.

At the same time, the contamination of products such as milk, meat and eggs with the residuals of medications, especially antibiotics, is especially common in more specialized livestock keeping, because of the indiscriminate and inadequate use of the chemical inputs. There is still a lack of knowledge and control in this respect (WHO, 2005).

References and Further Reading

FAO (2007) *The State of the World's Animal Genetic Resources for Food and Agriculture*. FAO, Rome.

FAO (2009) *The State of Food and Agriculture – Livestock in the Balance*. FAO, Rome.

Hooft, K. van't (2004) *Gracias a los Animales Crianza Pecuaria Familiar en America Latina, con casos de los Valles y el Altiplano de Bolivia (Thanks to the Animals – Family Level Livestock Keeping in Latin America with Case Studies from the Bolivian Valleys and Highlands)*. Plural Publishers, La Paz, Bolivia.

WHO (2005) The control of neglected zoonotic diseases – a route to poverty alleviation. Report of a Joint WHO/DFID-AHP Meeting. WHO, Geneva.

8

Changing from Smallholder Low-input and Diversified to More Specialized Livestock Keeping

_____

Learning Objectives: Understanding
- The background and reasons for changing from low input to more-specialized livestock keeping
- The characteristics of and requirements for the process of change
- The desired results and risks associated with changing from low-input strategies into more specialized systems

Reasons for Changing

As indicated in the previous chapters, a family or family member may decide to specialize the keeping system of usually one selected species whenever an opportunity presents itself. As depicted in Fig. 6.1 (page 54), the keeping of this species then changes from low input and diversified to a more specialized system. This implies a higher input level in terms of labour and other inputs than low-input systems. The rest of the animal species will remain under the low-input conditions.

When a family decides to dedicate itself more to one species and to intensify its husbandry, they may do so for various reasons. There are often special reasons within the family in combination with changes in the external situation. Some general factors may influence this transition, for example:

- Improved access to markets and other commercial centres, like better roads;

- Possibility of project support;
- Improved productive opportunities, such as irrigation or new types of crops;
- Opportunities within the ecosystem of the zone, especially the forage and water availability;
- Change in the social system.

There may also be personal, family or community changes that support this transition, such as:

- Loss of work and income in another sector;
- Opening up of a restaurant or other business where leftovers can be used to support animal production;
- Increased availability of money and labour input, for example when a member of the family returns from migration with extra cash;
- Increased community organization related to herding or breeding;
- The start of a producer association in the community;

- Increased level of education of one the family members, for example after graduation from agricultural college;
- Increased personal level of experimentation and self-learning;
- Change in land tenancy.

Selection of Species of Animal

The species of animal utilized to initiate more specialized husbandry depends on many factors, which may explain why there is a lot of variation among family strategies in spite of similar conditions and ecosystems:

- Preferences, culture and prior experiences of the family or organization to which they belong;
- Family preference for a species already known;
- Manual labour and quantity of land available within a family, to correspond with the need of the species to be selected for specialized husbandry;
- The area's climate and the availability of feed and water;
- External possibilities, such as a market to buy and sell products, transportation conditions as well as the possibilities and conditions of credit and technical services;
- Following the examples of other families;
- The offering of a project.

Various Examples

Examples of families changing the management of one species from low-input to a more specialized animal system are manifold. Here are some real-life examples from Bolivia, Mexico and Thailand.

- Vladimir in the Bolivian valleys used to work with his younger brothers in a small welding business. Some years ago, he started having problems with his eyes and he started to look for a possible alternative source of income. One day he met a Canadian religious man who wanted to support local people to set up small pig

businesses in the area. Vladimir started as the only one interested and now he has a specialized pig husbandry unit beside his house. His parents help him to take care of the animals (van't Hooft, 2004).
- The father of the Franco family in the Bolivian valleys has been one of the thousands of migrants that left his community 8 years ago to go to the large city for work. When he worked on the road, he was bitten by a snake from which he miraculously survived. Since then, he decided to stay and work at home with the family. Together with his wife and children, he decided to specialize in milk cows, making use of the cooperative milking group that had been set up in the community. With support of a dairy project, they started up their dairy. Three years later, they have a dairy with some 12 dairy cows, and additional income from buying and slaughtering cattle and selling the meat in different places (van't Hooft, 2004).
- The Garcia family in the Bolivian valleys had some cows raised under diversified husbandry by Mrs Garcia, while her husband worked on the road in another region. Mr Garcia always wanted to change to specialized dairy production because the milk market was affiliated with a dairy module that guaranteed the sale and price of the milk (Fig. 8.1). Nevertheless, for many years, his wife was not excited until the day when they went to visit a specialized dairy producer. Since that day, she decided that she was going to achieve the same at her house and together they have been able to build a specialized dairy in a few years (van't Hooft, 2004).
- In a small village in the Bolivian valleys, the Ochoa family has had a small restaurant for several years, where they also regularly sell homemade maize beer, known as 'chicha'. With the leftovers of the restaurant and the chicha making they grow pigs, which after slaughtering are served in the restaurant. They also regularly buy pigs on the market for this purpose. When a modern pig-raising farm was started nearby,

they decided to change from local breed pigs to pigs of the exogenous breed from this farm, on the one hand because these animals grow faster than the local breed animals and on the other to reduce the loss of meat of the pigs they buy on the market because of cysticercosis cysts (van't Hooft, 2004).

- In the hills of south central Mexico, a social outreach project of the Methodist Church of Mexico called Give Ye Them to Eat (GYTTE, local NGO) has been working since 1977 to combat hunger and poverty of the rural sector with the Heifer Passing on the Gift programme (Fig. 8.2). Their purpose is to strengthen the capabilities of marginalized people and communities to meet their basic needs. Using livestock as a tool for change, farmers are taught animal husbandry skills and animal nutrition practices for raising animals of exotic breeds of pigs and goats. Markets are set up for the improved inputs and produce coming into and out of the villages around the projects. A native pig that produces two to four piglets after a year is replaced with a crossbred type of pig that matures to piglet-bearing age within her first year and provides 8–12 piglets, with the capability of farrowing twice each year. The initial gift of livestock and training helps the first group of families to feed themselves as well as provide them with resources for future production. Then they go on to assist their neighbours by providing them with the same gifts that enabled them to improve their lives (Wollen, 2011, personal communication).

Fig. 8.1. This family in the Bolivian valleys is successfully keeping dairy cows in a more specialized way, which prevents out-migration and provides for the education of the children.

Fig. 8.2. Goats are handed to a woman in a Nepalese village through the Passing on the Gift programme of Heifer International.

- Damrong Taweesuksatit is a farmer in Chaing Dao District in the hills of northern Thailand, where poverty is high because of the high number of migrants, the steep hillsides and soil erosion during the rainy season. Damrong's village grows maize for human consumption as well as for pig production. The village residents favour pork, especially from black pigs. The common practice for animal nutrition is to harvest maize from the fields and to haul the grain to the city for grinding and then have the mill make a complete feed mix for the pigs, which has to be hauled back up into the hills to the village farms. With village ingenuity, Damrong was able to secure a small loan to purchase an electric grain grinder for the maize and a supply of mineral, vitamin and protein pig supplement to mix with the ground grain. With this, he was able to make the complete grain mix for the pigs, right there in the village. Damrong's business has flourished and has drawn additional business from outlying smaller communities. Because of this, many farmers have started rearing exotic breeds of pig. Even though there is greater risk attached to rearing these breeds of pig, the nutritional risks and costs are now minimized because of the complete feeding rations formulated in their own village (Wollen, 2011, personal communication).

The Desired Process of Change

Once the species has been selected, various elements are required to make the change. The elements vary by species and depend on the desire to work with crossbred animals or with purebred animals of specialized breed.

Capital

In order to initiate the change, it is necessary to borrow some money or invest capital obtained from migrations or other salaried activities at the time of initial investments such as the purchase of feed, animals, equipment and medications.

Animal feeding

Generally, the first thing that needs to be changed is animal feeding. The feeding must change in both quality and quantity with an adequate level throughout the year. With the improvement of feeding, body condition improves in local breed animals as well as in crosses and those of specialized breeds (Fig. 8.3). Individual animals begin to produce more and each one has a higher value.

Shelter

As one of the first steps, construction of an adequate shelter (Fig. 8.4) is necessary even though it is not necessary to think about very sophisticated construction. There are positive experiences with high-productivity animals in which simple construction made of locally available materials have been used and have been improved according to the growth in the quantity of animals.

Fig. 8.3. Feeding cows in a zero grazing system in Cambodia is a labour-intensive system, in which all feedstuffs must be brought to the animals. This gives positive results in terms of production, provided the feeding is continually guaranteed in quality and quantity.

Fig. 8.4. Animal shed made of local materials. The manure can easily be collected and used on the crop fields.

Purchase of animals

Often, the purchase of new animals is the beginning of the most visible change. With the purchase of animals of a specialized breed or crossbred animals, an increase in production can be achieved relatively quickly. At the same time, the investment and thus the risk is also higher. Specialization can also be achieved by utilizing a combination of improved management and a gradual change in the genetics of the local animals.

Veterinary support and disease control

With these changes, especially in the exotic and higher yielding breeds, there is a special need for other measures such as parasite control, vaccination and special care.

Professional organizations and associations of producers

The market factor and transport costs are part of the analysis of opportunities that must be made before initiating specialized husbandry. There are better possibilities of organizing the sale of products when agreements between families are reached (Fig. 8.5). An organization of families that dedicates itself to the same activity can achieve favourable mechanisms of transport, purchase and sale

of products, technical support and exchange of knowledge and experience.

Added value – value chain

In addition, the value of the products can be increased by preparing them at home and making secondary products that do not require immediate sale, such as cheese and leather crafts. In other cases, support is given to a larger investment to stimulate the value chain, such as in the case of the special slaughterhouse for llama meat in the Bolivian highlands (Fig. 8.6).

Technical support

During a period of change, families are particularly vulnerable. Knowledge required for the new method of husbandry is different from the knowledge required for diversified husbandry. In specialized husbandry, new diseases and risks are presented. Because of this, appropriate technical support is critical, especially during the first few years of the change, which is when there is much insecurity, unknown problems and debt that has no room for losses. Technical support does not necessarily have to come from professional technicians. More so, the advocates of animal health are of vital importance as well as neighbours and family members that have been through the same process of change.

Fig. 8.5. A dairy cooperative society in the Bolivian valleys supports smallholder families in marketing their milk. Being part of an organized group also facilitates other joint activities, such as centralized purchase of cottonseed and other products to reduce the costs of dry-season feeding.

Fig. 8.6. A llama slaughterhouse was part of an integrated support project to llama growers in the highlands of Bolivia. Through these slaughterhouses, it was possible to reduce the impact of the cysticercosis parasite in the meat, which helped to increase meat quality and price.

Positive and Negative Results

There are many projects whose objective is the change of diversified husbandry to more specialized husbandry. Moving to more specialized systems is in itself not a guarantee of success.

The change is considered positive when the family has stabilized itself with the new form of production, they express themselves positively with respect to the change and they succeed in improving their income and level of life. Conversely, the change is considered negative when the family has not been able to stabilize itself with the new productive form, they express themselves negatively with respect to the change and they have not achieved higher monetary income.

If adverse situations do not present themselves, the process of changing from low-input diversified animal keeping to more specialized animal keeping usually results in improvements of the economic situation of the family and produces social changes resulting from this increase in monetary income (Fig. 8.7). Thus, they can pay off their initial debts and continue specializing their animal husbandry.

The case of two families in the Bolivian valleys – the outcome of changing to more-specialized animal husbandry (van't Hooft, 2004) – are given below.

- Two years ago, Matilde and her family decided to specialize in dairy production. They bought two new cows and they cared for them intensively with their other three cows, dedicating special care to their nourishment. The children helped every day, and the husband who works on the road during the day has built a rustic roof and a special feeding area made from adobe (mud wall).

Fig. 8.7. The change of low-input to more specialized livestock keeping can lead to positive results at family level. At the same time, the risks of this strategy are considerable and need to be taken into account when designing a project in this direction.

The cows are now each producing double the quantity of milk that the other animals used to produce. The family has become one of the most important milk producers of the local dairy cooperative. Both the husband and wife are happy and proud of these results. Last year, Matilde was part of the cooperative leadership as the only woman in that position.

- Julio has wanted to specialize in dairy production over the past 2 years. He made silage, and with this, he began to improve the nourishment of his dairy cattle. The results were not as positive as he had expected because he did not have any able children at home to help him, and his wife did not have enough time for the cows because they had adopted an infant girl. The following year, Julio wanted to buy cottonseed to feed the cows during the dry season, but he could not agree with the other families on bringing a truckload from Santa Cruz. The cows were left without adequate nourishment and now Julio is thinking about selling them and starting a new business or emigrating to the USA.

Risks Associated with the Change

As indicated in Chapter 2, many livestock projects fail. The most frequent reasons are: over-ambitious project objectives, not including women, top-down methodologies, lack of practical experience, lack of social and cultural sensitivity, power differences in managing funds, failing communal livestock projects, failing marketing schemes and the introduction of exotic breeds in non-optimal conditions (Livestock in Development, 1999).

Because of these frequent and painful experiences, extreme and special care must be taken when starting up a livestock improvement project in which the aim is to change from low-input to more specialized livestock keeping. At a family level, account must be taken of the risks of this process, especially during the first few years. Elements such as initial investments, new diseases of the animals, and instability and dependence on external resources can cause the process to fail and leave the family in worse conditions than those they had before they began the project.

Appropriate technical support is thus critical, especially during the first few years of the change, which is when there is much insecurity, unknown problems and debt that leaves no room for losses. Technical support does not necessarily have to come from professional technicians. More so, the local animal health specialists, such as trained community animal health workers, are of vital importance – as well as neighbours and family members that have been through the same process of change.

Financial Risk

The availability of money is crucial, especially at the first stages in the process of change. Because of being dedicated to this new type of husbandry, there may be less income from other activities, while the income from the new activity may not yet have been generated, but borrowing more money from the bank at high interest represents a risk that many families with scarce resources do not always want to take.

Increased dependency and changes in the market

Families with more specialized animal keeping are increasingly dependent on the market on which they often have no control. For example, in Bolivia, an Angora rabbit project was started early in the 1990s that promised to give good financial results to smallholder farmers. This was positive for a couple of years until the market was taken over by Chinese farmers who could produce at lower costs. The project broke down and left the farmers with considerable losses as their investments in cages and other valuable equipment was not compensated (van't Hooft, 2004).

Unfamiliarity with new animal keeping systems

Knowledge required for more specialized husbandry is different from the knowledge required for diversified husbandry. During the period of change, families are particularly vulnerable. In more specialized systems, the diseases that affect animals are different from those in low-input systems, like for example milk fever in dairy cattle (Box 8.1).

Difficulty breeding high-producing animals and raising young stock

Most livestock projects that aim to improve productivity build on bringing in exotic livestock breeds to do so, in the form of crossbreeding, artificial insemination or bringing in live animals.

Many high-yielding animals have difficulties to come into oestrus and to breed when faced with feeding inadequacies (Fig. 8.8). This situation is worsened by common chronic uterine infections that often remain undetected and untreated. As a result, under local conditions of suboptimal nourishment and management, many of these potentially high-yielding animals fail to breed once a year; often breeding only once every 2 years. Over time, this often turns out worse than most local breeds would perform in terms of reproduction (LPP, LIFE network, IUCN-WISP and FAO, 2010).

The offspring of exotic and high-yielding breeds require special feeding conditions in order to grow and mature into healthy, productive animals. Faced with inadequate feeding during the dry season in combination with parasites and other diseases, these animals experience more difficulties in maturing normally than animals of local breeds. In many cases, it is difficult for families to keep up these conditions throughout the year.

Unexpected difficulties related to exotic breeds

Exotic breeds, when introduced without appropriate trials, can lead to unexpected difficulties. For example, in the humid tropics of Bolivia, the introduction of white pigs, such as Yorkshire and Landrace breeds, has not worked out. One of the adverse factors has been the existence of vampire bats

---

Box 8.1. Mortality from milk fever.

In specialized dairy production, there is a potentially lethal condition called milk fever or hypocalcaemia, which affects high-producing cows that have recently given birth. This is almost unknown to families that are only familiar with lower-producing cows.

Thus, in many dairy projects, the best dairy cows have died in the hands of families that had changed to this form of dairy production because of their unfamiliarity with this condition. Milk fever can be prevented by having a better understanding of nutrition. It can also be treated relatively easily if adequate measures are taken immediately when the problem arises, by infusing a calcium solution into the vein of the sick animal by a trained person. This requires special attention and skills at the time of parturition (van't Hooft, 2004).

in the zone, which attack at night to suck blood (Fig. 8.9). From this, young female pigs may lose their nipples before the first birth. This often goes unnoticed. As a consequence, the animals lose their ability to feed their young, which die shortly after birth. This problem does not affect the local breeds with red or black skin (van't Hooft, 2004).

Long-term effects on biodiversity and environment

Efforts to increase the income of poor rural farmers through micro-credit are often directly or indirectly linked to improved animal health practices within more specialized systems. Those results may be positive in the beginning, yet they could have serious negative side-effects in the long-run. Many of these programmes introduce Green Revolution-type technologies, such as cross-breeding with exceptionally high-producing breeds, use of an abundance of commercial fertilizer, improved seeds that require additional care and other commercial inputs. Without proper training and persistently improved management, these measures have often led to serious environmental degradation, loss of locally valuable genetic breeds and high vulnerability to financial obligations beyond the ability to repay for the families involved.

Fig. 8.8. In the case of inadequate feeding, especially the exotic and high-producing animals – both purebreeds and crossbreeds – do not perform well and experience difficulties with repeated breeding, poor calf-growth and mortality.

Fig. 8.9. Exotic breeds, such as these light-coloured pigs, can experience unexpected difficulties in the harsh environment of the tropics, for example because of attacks from vampire bats.

References and Further Reading

Hooft, K. van't (2004) *Gracias a los Animales Crianza Pecuaria Familiar en America Latina, con casos de los Valles y el Altiplano de Bolivia (Thanks to the Animals – Family Level Livestock Keeping in Latin America with Case Studies from the Bolivian Valleys and Highlands)*. Plural Publishers, La Paz, Bolivia.

Livestock in Development (1999) *Livestock in Poverty-focused Development*. Livestock in Development, Crewkerne, Somerset, UK.

LPP, LIFE network, IUCN-WISP and FAO (2010) Adding value to livestock diversity. Marketing to promote local breeds and improve livelihoods. Animal Production and Health Paper, no. 168. FAO, Rome.

9

Recommendations for Optimizing Smallholder Low-input and Diversified Livestock Keeping

---

Learning Objectives: Understanding
- The need to adapt recommendations to livestock keeping strategy
- Main objective in supporting low-input and diversified livestock keeping: reducing mortality
- Main causes of animal mortality in low-input livestock keeping
- Recommendations for stimulating low-input systems in each of the eight special areas

Adapt Recommendations to Local Circumstances

It is important to analyse the type of animal husbandry used within a family or community, as well as the local circumstances at hand, before embarking on activities in support of the animal husbandry practices. Because of the differences in objectives between systems, as explained in the previous chapters, it is necessary similarly to adapt the recommendations. This will be done for low-input and diversified smallholder systems in this chapter, and for more specialized smallholder systems in Chapter 10. The aspects related to marketing will be detailed in Chapter 11.

Most livestock keepers in the world can be found in the low-input and diversified smallholder farmer and pastoralist systems. In addition to the low-input and diversified animal keeping of various species, the same family may be employing a more specialized type of animal production of one selected species.

Main Goal: Reducing Animal Mortality

In low-input and diversified husbandry, the animals are raised with minimal input in terms of labour and feeding costs. The animals generally roam around to find their own feeds – or may be fed with leftovers. It is accepted that the animals lose weight during the lean season. This is, however, quite an economically viable system, because all produce is direct gain, as no significant costs are needed.

Within this system, it is not possible to generate greater earnings by reducing production costs, because these costs are already minimal. Neither can per-animal productivity be substantially increased because this implies an increased investment of labour and capital, which goes against the logic of this type of husbandry.

Therefore, the best way of stimulating low-input and diversified husbandry within its own logic is to reduce the animal mortality (Fig. 9.1; van't Hooft, 2004).

Mortality in this type of husbandry can vary depending on the species, the climate, the season, management, and the presence of predators and the epidemics of infectious diseases. Mortality can be as high as 80–90% in the case of chickens, for example. Reducing the mortality of chicks from 80% to 60% for example, will double the quantity of live chicks. For this reason, families with diversified husbandry will dedicate effort to avoiding mortality, but rarely try to increase per capita animal production. In order to find the means to reduce mortality, we must therefore take into account the limitations that characterize this type of production.

For this reason, measures for reducing animal mortality should be inexpensive and employ little manual labour; the earnings achieved in the short term – in the form of live animals – must be greater than the costs necessary to achieve the change.

In order to improve low-input and diversified animal keeping, one needs to understand the main reasons for mortality. Generally speaking, besides culling for family needs, the eight main causes of mortality in low-input livestock keeping are:

1. Nutritional deficiencies, especially during dry periods;
2. Lack of pasture;
3. Water deficiencies;
4. Infectious diseases;
5. Internal and external parasites;
6. Breeding deficiencies;
7. Lack of protection;
8. Lack of care during special moments (birth, illness).

In this chapter, these eight elements will be used as the entry point for listing improvements at farm level, aiming at reduced animal mortality. Please note that this is an overview of possibilities of improved management practices rather than a complete guide (Fig. 9.2).

1: Improved Animal Nutrition

In many community livestock development projects, the objective of improving nourishment during the dry season is one of the main themes. Despite this, the results have not often been very encouraging. Very few times have the introduced improvements been adopted in large part by the families. This may be linked to the fact that changing the feeding strategy often implies a change from low-input to more specialized system.

Deficient nutrition during the dry and cold months of the year is in itself one of the characteristics of low-input and diversified husbandry. The objective of the improvement in low-input livestock keeping, therefore, is to reduce the mortality during the dry season rather than increasing animal productivity (Table 9.1).

The recommendations for animal nutrition in low-input livestock keeping are divided into two parts: (i) dry season nutrition; and (ii) mineral supply.

Fig. 9.1. Animal mortality is the major reason for loss in low-input livestock keeping. The recommendations for improvement are therefore based on reducing animal mortality. The costs of doing so need to be less than the gain of the animals saved.

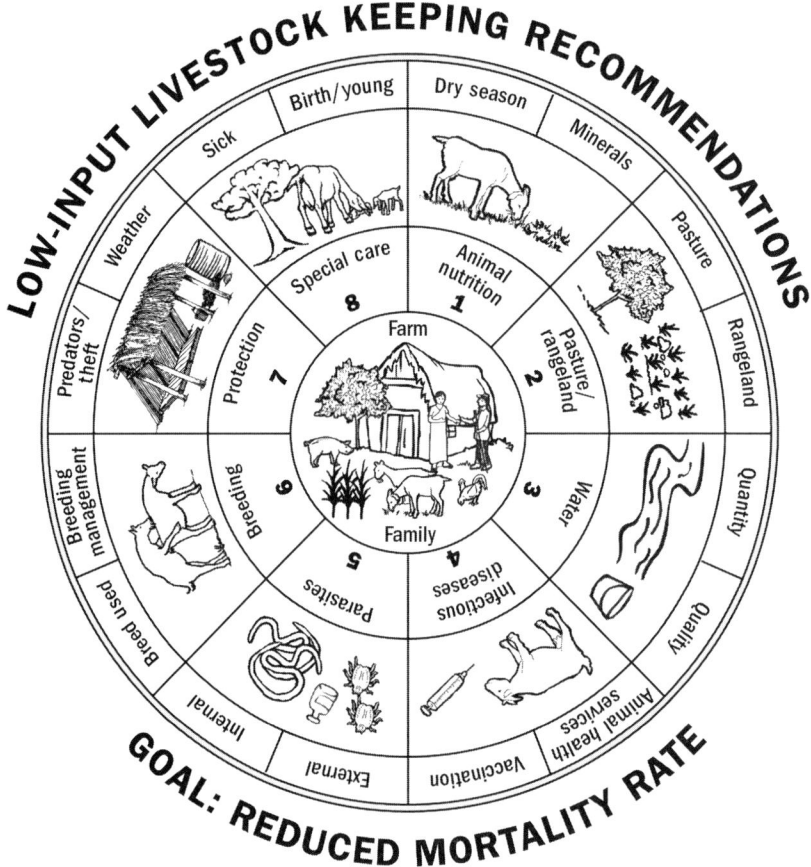

Fig. 9.2. The wheel of animal well-being and production for low-input livestock keeping. The recommendations to support low-input and diversified livestock keeping are based on the eight major reasons for mortality.

Table 9.1. Objectives and recommendations for improved animal nutrition in smallholder low-input and diversified livestock keeping.

Animal nutrition	Objectives	Dry season nutrition	Mineral supply
Low-input and diversified systems	Reduced mortality in dry season Reduced weight-loss Increased resistance to drought		
Recommendations for improvements		Agriculture leftovers storage and feeding Support local feeding innovations Plant leguminous trees Improved use of kitchen leftovers Green forage Hay making Cheap by-products Feed troughs	Provide ordinary salt or home-made mineral blocks Vitamins

Dry season nutrition

In low-input and diversified livestock keeping, the lack of food during the cold or dry season is in itself the cause of a sizeable proportion of mortality. The process for improving nourishment without leaving the logic of diversified husbandry is very complicated, as improving animal feeding generally requires significant investment of time and money.

The best way of improving nourishment in this kind of animal husbandry without abandoning the logic of little investment seems to be supporting the traditional ways of animal feeding and all other local innovations that are accepted and adapted to the local circumstances. The feed needs to be improved in both quality and quantity. In this process of improvement, the characteristics and limitations of each species must be taken into account. Moreover, the way of presenting the feed can be improved in simple ways.

Improved straw storage and feeding

In most tropical climates around the world, there are traditional forms of procuring and saving animal feed out of crop residues, such as maize husks. The residues depend upon the crops grown on the farms; the residues of rice, wheat, millet or maize straw are major animal forages. In low-input diversified animal keeping, stover is usually handled and dried in the long, unchopped state (Fig. 9.3) and stored in varieties of ways. In addition, groundnut tops, halms of peas, cowpeas and soybean are also used for off-season feeding.

In many cases, the traditional storage methods or the provision of such feeds can be improved, for example by reducing the time of drying the stalks. Better results are also achieved when these feeds are piled in the shade, as they maintain more of their nutrients this way.

These feeds are low in nutrients and minerals and for this reason, it is best to chop into smaller pieces and to combine with salt. These methods can enhance the nutritional value of these feeds and make it more palatable:

Fig. 9.3. Rice straw is sun-dried, transported and stored for off-season feeding to cattle and buffaloes in Nepal.

- Cut the straw into 1–2-inch long pieces
- Put it into a bucket or big vessel
- Cover the straw with water
- Salt can also be added as required
- It can be fed after a few hours
- Green grasses can also be added to make it tastier.

*Supporting local feeding innovations
and traditions*

In the Andean highlands, some farmers are using an aquatic plant locally known as q'hora for feeding their animals. Some farmers make hay from this plant. Such local possibilities can be researched and further explored to optimize animal keeping without major costs. Another example of a tradition that can be supported is to plant certain leguminous tree species, which do not only provide shade and live-fences, but also nutritious oily seeds during the period of highest feed shortage. A third example of a local innovation from the Bolivian valleys is the use of chicken manure to provide some extra nutrients for sheep during the dry season, as an alternative, inexpensive and nutrient-dense feed source (van't Hooft, 2004).

Improved use of kitchen leftovers

Pigs and other small animals are often fed on leftovers from home kitchens, restaurants or local breweries. In general, cooking the left-overs and adding salt in the case of raw products will increase the digestibility.

Green forage

Some smallholder farmers grow small quantities of specific crops to feed their animals during the dry season. Examples are oats and barley. Oats yield more and are more palatable than barley. In high-altitude areas, barley is cultivated more than oats for its yield capacity in less fertile soils and for its ease of transport and storage. Other farmers grow lucerne – a high quality legume – to feed their animals throughout the year. Different varieties of lucerne are used depending on the climatic and soil conditions in each area. Lucerne is a more difficult crop, as it requires irrigation during the dry months. Because of this, in some areas it is sold as a cash crop.

Hay making

Some farmers produce hay and store for the dry season. This is not very familiar in low-input animal keeping systems, however, and attempts to introduce it in regions where the custom does not exist often tend to fail. Hay can be made from various crops such as barley, oats or lucerne. To produce it, the crop is allowed to dry for 1.5–2 days. Care should be taken not to dry it for too long, as nutrients are lost in the process. There are different ways to store the hay after the drying process. Dried lucerne or lucerne hay is a good-quality product, but making lucerne hay requires manual labour during the very busy rainy season, so many prefer not to undertake this endeavour.

Cheap and easy to obtain by-products

Cheap by-products are often commonly used in low-input and diversified livestock keeping. An example is the common use of wheat and rice bran in Bolivia, a by-product from community wheat and rice mills that can be purchased in small quantities. The bran is a feed rich in vitamins and proteins and is often used as a by-product to feed pigs, horses and cattle. This can be improved by adding salt and the right quantities of water.

Use of feeder troughs

The quality and quantity of feed can also be improved by using and improving the feeder troughs used. A trough without holes and of sufficient size can avoid the loss of feed and reduce the incidence of internal parasites.

Salt and other minerals

Provide kitchen salt

For all animal species of all ages and in all climates, supplying minerals is one of the most important elements in low-input diversified husbandry. At a minimum, common salt (sodium chloride) is required, like that found in kitchen salt. It is best to provide it in small quantities every day, even though it is also possible to ration it to livestock two times per week. These minerals facilitate the body in taking advantage of the little nourishment that it acquires so the animal can better process dry feed of poor quality. In this way, salt strengthens general health, resistance and reproduction. When ordinary salt is not given to the animals, they look for it by licking earth and sweat or biting bones. The cost of salt is relatively easy to recuperate, with the increase in milk production in the short term and the increase in growth of young stock and reproduction in adult stock in the medium term.

Preparation of simple mineral block

Ordinary salt is best given in combination with other minerals. Different types of mineral block can be prepared easily in the field depending upon the availability of the resources (Fig. 9.4).

METHOD OF MINERAL BLOCK PREPARATION (ANIMAL HEALTH TRAINING AND CONSULTANCY SERVICE, AHTCS)

Materials:

- ½ kg red (iron rich) soil
- ½ kg common salt

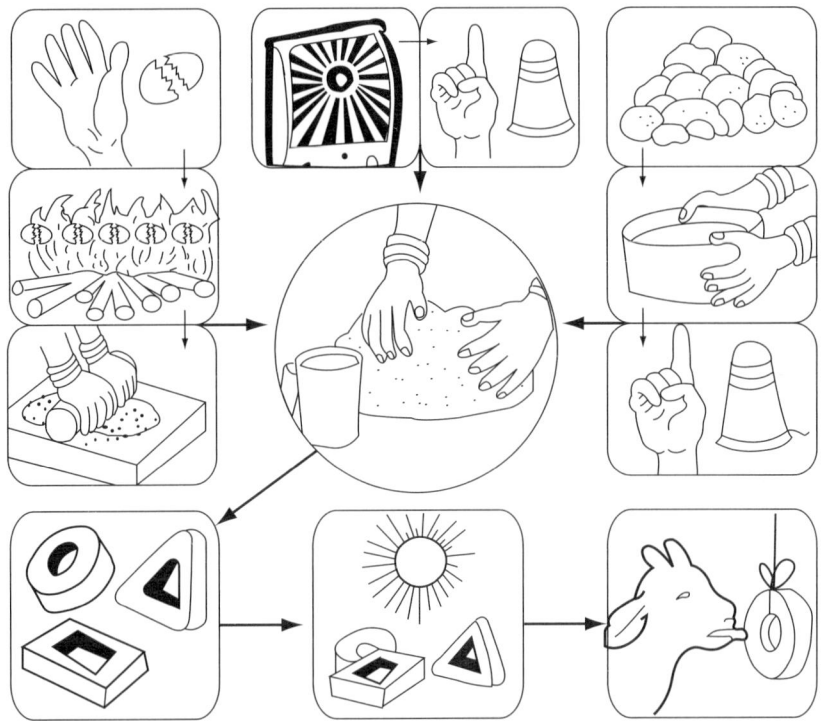

Fig. 9.4. The preparation of a mineral block in specific countries in South Asia and Africa from locally available materials: red soil, common salt, eggshells and a handful of wheat flour. After elaboration, the block is placed in the shelter where the animals spend the night. Credit: Bhandari (2009).

- 5 eggshells (shells only)
- 1 handful flour (wheat or otherwise)

Method:

- Heat/roast eggshells and crush into fine powder
- Grind the soil and salt into a fine powder
- Mix dry ingredients with water to form a paste; mould into desired shape
- Form a hole in middle of shape in order to hang block with a string from overhead
- Allow block to dry thoroughly; no sunlight for 2 full days so that it dries from the inside; then in sunlight to complete drying

Hang the block:

- Acclimatize animals to the block by putting it in the feed box for a day or two
- Hang in stall and under cover so that it does not erode in the rain
- Keep mineral block available at all times

Another example from northern Ghana shows that natural mineral licks can be made from special salty soil. Local farmers experimented with adding other nutrients to this soil, especially oyster shell and cassava flour, with support from the CSIR Animal Research Institute in Tamale, Ghana (Karbo, 1999).

Vitamins

There are several essential vitamins that animals need, especially during the dry season. Vitamins have functions similar to those of minerals. There are differences between species with respect to needs in different periods. While ruminants can synthesize the majority of the vitamins in the rumen, vitamins need to be supplied to non-ruminants. Although there is no clarity about the necessary quantities of different vitamins in low-input animal husbandry, the use of inexpensive multi-vitamin products is generally recommended.

2: Improved Pasture and Rangeland Management

Overgrazing is a common problem associated with low-input and diversified livestock keeping, especially with ruminants in dryland and rain-fed areas. Goats are perceived as especially destructive in this perspective. Overgrazing is an initial process, which triggers land degradation and favours further poverty of soils, plants and people. The problem is often especially acute in communal grazing areas. At the same time, when managed correctly, animals can also provide the key to improving soil fertility and forage production (Table 9.2).

The recommendations for pasture and rangeland management in low-input and diversified livestock keeping are divided into two parts: (i) pasture management; and (ii) rangeland management.

Improved pasture management

Controlled grazing instead of roaming animals

Many families strive for pasture management by pinning individual animals on a specific spot during a few hours or by herding (Fig. 9.5). This is often used for sheep and goats as well as pigs, and represents an efficient way of controlled grazing.

Weed control

A certain level of critical weed control is often taken care of by strategic rotation of animal species or mechanical means.

Zero-grazing system

In order to reduce the pressure on the pastures, sometimes it is advantageous to use the zero-grazing system, in which the feed

Table 9.2. Objectives and recommendations for improved pasture and rangeland management in smallholder low-input and diversified livestock keeping.

Pasture and rangeland management	Objectives	Pasture	Rangeland
Low-input and diversified systems	Reduced overgrazing and soil erosion Reduced brush encroachment Increased carrying capacity Increased resistance against drought Community organization		
Recommendations for improvements		Controlled grazing Zero-grazing systems	Reviving communal grazing control Fence off grazing areas Rotational grazing Special grazing areas for dry period Controlled and prescribed fire

Fig. 9.5. Herding is a common way of controlled grazing and pasture management. This also limits the opportunity of youngsters to attend school.

is brought to the animals. This is often recommended in the case of goats. In most cases, however, this implies a transition to a more specialized animal keeping system.

Improved rangeland management

In low-input and diversified systems, rangeland management objectives relate to reducing the encroachment of brush, reducing soil erosion, increasing the carrying capacity of the land and managing through drought conditions. This is done by using long-standing practices of community-managed rotation of herds and flocks. Indigenous people have recognized ways of preparing for droughts, prevention of overgrazing and protecting fragile areas for many years within their rangelands (Flintan and Cullis, 2010).

Reviving communal grazing control

A workable plan for regulating the grazing of communal grasslands in diversified animal raising communities is to re-establish traditional structures and systems. For example, in the most isolated zones of Bolivia, these structures are still in force today. Their communal decisions are based on climatic predictions and a profound knowledge of the productive capacity of the local forages. At the same time, there are continuous discussions on the need to maintain the animal load in balance with the capacity of the grasslands. There are large variations according to zone and it is very difficult to determine exactly the optimal animal load for each zone (Fig. 9.6).

Use of traditional animal species

Animal species such as camels, llamas and alpacas are less destructive for fragile pasturelands than other ruminant species, such as cattle, goats and sheep. The camelid species have soft-padded feet that prevent soil erosion. Moreover, their feeding habits can actually stimulate the growth of palatable plant species.

Fig. 9.6. Alpaca grazing in the highlands of Peru. The conditions of pastures and rangelands are very diverse, which makes it very difficult to generalize about the carrying capacity of this resource.

Fencing off grazing areas

Traditional fences made of stone can be found especially in the hilly areas. Other fences are made with posts and barbed wire. Some projects with camelids in the Altiplano of Bolivia aim to create native fields of protected grazing, delineating certain special grazing areas with such fences. The delineation has the advantages of improving management of native grasslands and freeing up the shepherd to do other things. At the same time, there may well be disadvantages to this activity, as it interrupts the traditional movement of animals, which may well have an important ecological rationale (van't Hooft, 2004).

Rotational grazing

Rotational grazing practices, either with herders or fencing, can improve the utilization of local forages or extend the grazing season for all livestock. Rotational grazing is periodically moving livestock to fresh paddocks, which allows pastures to regrow. Rotational grazing requires close monitoring. If used properly, animal health improves and soil nutrients are effectively managed.

Reviving indigenous practices to reduce bush encroachment

Efforts by pastoral communities to revive indigenous rangeland management practices, such as the use of prescribed fire, are now gaining the attention of policy makers. For example, in the Borana Plateau in southern Ethiopia, heavy grazing by livestock, reduced mobility of pastoralists and lack of fire have contributed to the conversion of open mixed savannah to dense woodlands and bush lands. Herbaceous forage production for cattle and sheep has declined because of competition with woody plants for water and light. Residual grass is subjected to intense grazing pressure, further exacerbating the downward spiral.

Prescribed fire is the traditional most cost-effective means of manipulating vegetation in the savannah ecosystems of Eastern Africa. However, a blanket ban on the use of fire was introduced during the 1970s. This was intended to protect cropland and other natural resources from indiscriminate burning. An unintended consequence of this policy has been a weakening of traditional forms of range management that depended, in part, on the regulated use of fire to control

Fig. 9.7. Fire applied in a rangeland area in a controlled way in southern Ethiopia, to control undesirable woody plants, to promote forage production and to reduce the incidence of disease-carrying ticks. Credit: Getachew Gebru Tegegn.

undesirable woody plants, to promote herbaceous forage production and to reduce populations of disease-carrying ticks. An alliance of pastoral communities, researchers, policy makers and development actors is now experimenting with ways to re-introduce prescribed burning into the Borana Plateau (Fig. 9.7). They have found that useful trees, such as the acacia, remain intact after the prescribed fire (Gebru Tegegn, 2010).

3: Water Provision

Dirty or insufficient water is a factor that seriously limits low-input and diversified animal husbandry. This essential and basic element is too often neglected and under-estimated, especially in smallholder conditions. The possibilities logically depend on the conditions of each place (Table 9.3).

The recommendations for water provision in low-input livestock keeping are divided into two parts: (i) access to water; and (ii) water quality.

Access to water

Regular watering

In low-input animal keeping, the access to water is usually limited, as this often requires special resources and inputs. Moreover, water can also be a vehicle for diseases to pass from animals to humans (zoonoses). This can be a major and often under-estimated reason for animal as well as human disease and mortality. Depending on the species, animals need to be watered at least once or twice a day. With sufficient clean water – and especially in combination with sufficient minerals (especially common salt) – the animals can much better resist the periods in which they receive limited nourishment.

Opt for animal species or breeds that require less water

Some animal species, especially camelids, require very little water in comparison to other species. This is especially relevant in

Table 9.3. Objectives and recommendations for improved water provision in smallholder low-input and diversified livestock keeping.

Water	Objectives	Access to water	Water quality
Low-input and diversified systems	Regular water uptake Water quality sufficient Pollution of human water-source prevented		
Recommendations for improvements		Water 1–2 times per day	Prevent polluted water sources
		Animal species that require less water	Protect human water supplies

dryland and desert areas. For example, in the deserts of Rajasthan in northern India, camels are better suited than cattle. They can subsist entirely on the local drought-resistant trees. Moreover, their soft-padded feet prevent soil erosion and their feeding habits can stimulate the growth of desert plants (Köhler-Rollefson, 2010).

Water quality

Prevent polluted drinking water for animals

Drinking water can be of insufficient quality, related to various factors:

- Muddy water source in which animals have trampled to get to the water;
- Non-cleaning of the water resource, such as cans or tins;
- Water used has been polluted, for example with soaps, chemicals or residues, for example, from mining activities.

Prevent pollution of drinking water for human consumption by animals

In low-input animal keeping, people and animals live closely together. Animals can pollute the drinking water for humans, which can be a source of disease for humans. This can happen for example:

- When they can defecate in the source of drinking water for humans;
- When animals drink from cans that are later used for human consumption;

- When dead animals are present in swimming or drinking water.

4: Control of Infectious Diseases

Contagious infectious diseases with high mortality are a common problem in low-input and diversified livestock keeping. Each animal species has one or two major infectious diseases that are often possible to prevent relatively easy. In order to accomplish the aim of reducing animal mortality, it is necessary to focus especially on the control of these infectious diseases (Table 9.4).

The recommendations for the control of infectious diseases in low-input livestock keeping are divided into two parts: (i) animal health services; and (ii) vaccination.

Animal health services

Animal health services usually aim at a combination of disease treatment and disease prevention.

Supporting ethno-veterinary practices and practitioners

To keep animals healthy, traditional healing and prevention practices have been applied for centuries, and have been passed down orally from generation to generation. Local health care systems combine the application of medicinal plants with careful grazing,

Table 9.4. Objectives and recommendations for improved control of infectious diseases in smallholder low-input and diversified livestock keeping.

Infectious diseases	Objectives	Animal health services	Vaccination
Low-input and diversified systems	Reduced incidence zoonosis Reduced animal mortality because of infectious disease Promote synergy between traditional and modern remedies Improved access to local animal health services		
Recommendations for improvements		Support ethno-vet practices and practitioners Train community animal health workers Awareness of zoonosis	Vaccination of 1 or 2 major infectious diseases

feeding and breeding management. These so-called 'ethno-veterinary practices' reflect the in-depth knowledge of livestock keepers of their animals, ways to prevent and cure livestock diseases, as well as their environment.

It is becoming more common now for ethno-veterinary practices and collections of remedies to be recorded and even published. The wealth of information and methods to use ethno-veterinary medicines are collected, more commonly, by a few people in each community (Fig. 9.8). These healers are respected and valuable people for the health care of the animals as well as the people of the region (FRLHT and Tanuvas University, 2010).

Promote synergy between traditional and modern remedies

To support low-input animal husbandry, it has become increasingly recognized that enhancing ethno-veterinary practices and supporting the community animal health workers can play an essential role. This can lead to synergy between the use of local and modern remedies. The modern remedies most used in these systems are vaccines and very few antibiotics. These are of special

Fig. 9.8. The indigenous Tzotzil women pastoralists in Chiapas in southern Mexico have learned to prevent and cure the diseases of their animals, especially in their sheep. Credit: Ellen Geerlings.

importance for acute problems, especially in the case of species that involve cash (e.g. cattle and pigs). In most other cases, traditional remedies are used in several parts of the world. Community animal health workers provide basic modern remedies and they also use only effective local remedies to treat several diseases.

Train Community Animal Health Workers (CAHW)

In regions with low-input livestock keeping, animal health care is usually left up to livestock keepers. There is no private veterinary practice available to provide health advice, services or product – and if there is, the costs are usually far too high. The government livestock veterinarians are often engaged in administrative issues at regional and national levels. Livestock keepers are left with their own ethno-vet experience to protect and treat illness and injury.

Throughout the world, there are programmes to train community animal health workers: these are specially trained local community members who help fellow-farmers and community groups to prevent and cure animal diseases, and to optimize animal production (Fig. 9.9). The primary purpose of the CAHW programme is to increase the access to affordable, basic, animal health services. The sustainability of community-based animal health care is enhanced with payment for service by the farmers and by providing adequate training to the workers. Often the community animal health workers combine traditional and modern medicine, but this is not always the case. They will especially favour modern medicine when this is their major source of income.

Increase awareness about zoonosis

Zoonoses are diseases that are passed from animals to humans. These diseases are especially common in low-input livestock keeping, because of the close ways of living together. Several of the zoonoses are officially considered neglected by the World Health Organization (WHO, 2005). One of these, a very common parasitic disease in pigs, is cysticercosis.

Fig. 9.9. Community-based animal health worker in the Philippines prepares a treatment. These trained local famers provide cheap animal health services at the doorstep. The activities of these practitioners are not officially recognized in all countries, however.

Cysticercosis is found as cysts in the flesh of pigs and cattle. The parasite can be passed to people via raw or poorly cooked meat, and will result in long tapeworms (*Taenia* species) in the intestines of people. While training community animal health workers, it is also good to inform them about the ways to prevent this zoonotic disease – in the case of cysticercosis, by keeping pigs away from human stools, and by thoroughly cooking all pig meat before consumption.

Disease prevention through vaccination

Many livestock species have one or two infectious diseases that cause occasional high levels of mortality. The necessary vaccinations depend on the major diseases prevalent in each zone or country, for example in Latin America, swine fever in the case of pigs; Newcastle disease in the case of chickens; blackleg, hemorrhagic septicaemia and anthrax. In some cases, for example Newcastle disease in chickens, the occasional outbreak can cause the death of up to 90% of the animals. The other major reason for vaccination is to prevent zoonoses, such as rabies in dogs and cattle.

Vaccination against infectious diseases with high animal mortality

Modern vaccination is one of the main ways to reduce mortality in diversified husbandry, as it has been tested in different projects in different countries around the world. The combination of modern vaccination opportunities with traditional forms of treating and avoiding these diseases is of great importance.

There are, however, various factors that complicate the practice of vaccination in diversified husbandry. Most vaccines have been developed under the conditions of high-input and specialized husbandry. Some of the most common difficulties in promoting vaccination against major infectious diseases are:

- Vaccines requires a monetary investment when the animals are healthy,

which goes against the logic of this husbandry when there is no understanding of the importance and function of the vaccine. If the illness has not appeared over several years, families often choose not to vaccinate – even though the risk is greater than in the case of more recent outbreaks because the animals do not have a developed resistance.

- Vaccines are inexpensive and effective, but often come in large quantities and in single-dose bottles that must be used in a limited period. This is difficult for the conditions of diversified low-input husbandry.

- In order to be effective, the majority of vaccines require constant refrigeration from the time they are made until their application to the animal; this is complicated by the conditions of the farming areas in diversified husbandry.

- There can be side-effects with some vaccines.

- Because of the large quantities and the need for constant refrigeration, the application of the majority of the vaccines usually requires organization of campaigns between the families and the community animal health workers that apply them (Box 9.1).

Vaccination against infectious diseases that are required by law

In many countries, vaccination against certain diseases is obligatory by law – or strongly re-enforced by government regulations. These diseases are often of major importance for the export of animal products, especially meat to the EU or the USA where many diseases are strictly regulated. At the same time, the disease may not be a serious threat to low-input animal keeping, as it has become endemic. This means that most animals have developed antibodies against the disease and do not die of it. In many other cases, farmers have developed ethno-veterinary remedies to cure the animals.

An important example is foot-and-mouth disease. The fact that the disease does not seriously affect their animals

Box 9.1. Vaccination of birds against Newcastle disease in southern Nicaragua.

In the Project of Rural Integral Development 'Manuel Lopez' in El Sauce, Nicaragua, the vaccination programme against Newcastle disease was gradually developed, based on trained advocates, mostly women, for every 20–50 families. The day prior to the vaccination, the advocates were going to retrieve the vaccine for the project, returning to their houses with the vaccine in ice in a coolbox. In the early morning of the following day, the female participants received this vaccine in a 1-cc syringe with some ice cubes to be able later to vaccinate their animals that they had closed up for the night. A little drop was placed in one eye of each animal with the syringe without a needle (Fig. 9.10). In this way, approximately 120,000 birds were vaccinated two to three times a year (Kyvsgaard et al., 2001).

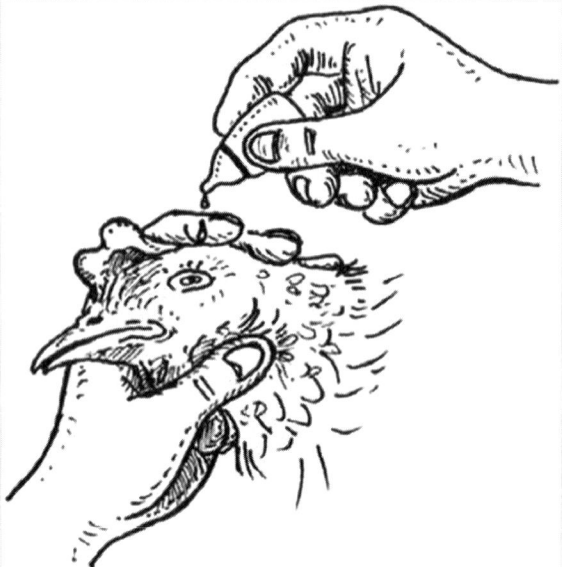

Fig. 9.10. Vaccinating local chickens against the common and highly contagious Newcastle disease can give good results, but requires good organization and trained animal health workers.

implies that farmers may not be motivated to spend effort and money on vaccinating their animals. This is worsened when there are (real or false) side-effects of the vaccine. Special incentives for vaccination may be necessary in this case – and it may be of national interest to do so.

5: Parasite Control

The reduction of parasite levels is another key element in low-input and diversified animal keeping. Usually there is multi-parasitism, with a large number of different internal and external parasites affecting the animals, especially the young ones. In general, parasites have a worse effect on young animals and exotic breeds – such as Holstein cows or Yorkshire pigs – than on adults and animals of local breeds. They also affect weak and malnourished animals more than healthy ones. Moreover, in warm and humid climates, the problem of parasites is worse than in dry and cold climates. Therefore, the incidence and the necessary controls have many elements that vary according to zone, species, breed, age and general state of the animals. For this reason, specific guidelines are difficult (Table 9.5).

Table 9.5. Objectives and recommendations for improved parasite control in smallholder low-input and diversified livestock keeping.

Parasite control	Objectives	Internal Parasites	External Parasites
Low-input and diversified systems	Reduced incidence of internal and external parasites Preventive parasitic zoonosis Reduced loss of young stock because of parasites Improved leather quality		
Recommendations for improvements		Make use of natural resistance of local breeds Reducing parasite incidence in grazing and feeding areas Parasite control especially in young stock Support ethno-vet remedies for parasite control	Make use of natural resistance of local breeds Use ethno-vet remedies for parasite control Community control activities (dip/baths)

Another problem related to internal parasites in animals is that some of them can pass to people (so-called parasitic zoonosis).

In general, it is worth the effort to explore the possibilities of controlling parasites within the logic of low-input animal keeping for each species and according to the characteristics of each zone. The objective of internal parasite control is not eradication of the parasites but rather control of their number while the animal can build up sufficient resistance. When the animal grows older, it can better deal with the parasites, at least under normal circumstances. For this purpose, both ethno-veterinary and commercial treatments can be used.

The recommendations of parasite control in low-input livestock keeping are divided into two parts: (i) internal parasite reduction; and (ii) external parasite reduction.

Internal parasite reduction

Most types of internal parasites reproduce through eggs that come from the animal with the manure. The eggs can survive outside the body for a certain period, and develop through several stages and life forms. Finally, the juvenile forms climb up the plants to be eaten by livestock. This process is greatly enhanced by high temperatures and humidity. As a result, the control of parasites in all animal keeping systems is a combination between animal-related measures and control of grazing.

*Make use of natural resistance
of local breeds*

Several local breeds have innate resistance against (external and internal) parasites. In traditional systems, this is consciously stimulated in selecting the breeding animals with lowest problems related to malnutrition and parasites. This system has been broken down by introducing animals of high-yielding exotic breeds that do not have natural resistance against most internal and external parasites. One of the ways to reduce the impact of parasites is to use local breeds again instead of exotic animals or crossbreeds.

*Reduce parasite incidence in grazing
and feeding areas*

Understanding parasite life cycles allows us
to manage the grazing areas of our livestock
and can be a means of avoiding their infective stages. Pasture rotation and the use of
feeding troughs is a part of preventative and
curative anti-parasitic management. Feeding
on the ground increases the risk of parasitic
disease.

Special focus on young animals

Internal parasites have a dramatic effect on
young animals, of both exotic and local
breeds – though the effect on young animals
of exotic breeds is more dramatic. The
young become thin, their bellies swell, they
do not grow normally and eventually
they may die. Young animals need extra
treatments to reduce parasite numbers
(Fig. 9.11) – whereas good nutrition can
help to overcome the effects of parasites to a
large extent.

Medications for parasite control can be
on basis of plant treatment (ethno-veterinary
treatments) or as commercial products. The
latter are usually more expensive. Therefore,
these commercial products are not often
used in low-input animal keeping, except
for large animals or in case of emergency.
The commercial products lead to more rigorous expulsion of internal parasites. These
products must be used with a thorough
understanding and compliance with the

Fig. 9.11. In young animals, regular drenching with
medicinal plants against internal parasites (ethno-
veterinary treatment) can largely reduce parasite
numbers.

label recommendations, as indiscriminate
use can lead to resistance of the parasites
against the drug.

*Support ethno-veterinary treatments
and practices for internal parasite control*

Traditional remedies for the control of different types of parasites can be stimulated.
Examples are:

• Regular drenching with medicinal
  plant liquids against internal parasites
  (ethno-vet treatment) to reduce parasite
  numbers – all ages but especially and
  regularly in young animals;
• Control grazing with muzzles while
  passing through areas with parasites,
  such as in the case of liver flukes in
  Chiapas, Mexico (Perezgrovas, 2006).

External parasite reduction

All animals can be hosts to external parasites that live on the skin, such as ticks, lice,
mites, fleas and flies. In llama and alpaca
husbandry, for example, external parasites
are one of the most serious problems.
Animals with mange and lice become thin
and skinny, scratch themselves and injure
their skin and the wool.

Ticks can also infect and transmit certain internal diseases such as redwater, gall
sickness and heartwater. Exotic breeds tend
to be at greater risk from these diseases than
indigenous and local breeds. Older animals
are more at risk than young animals for redwater and gall sickness.

External parasites, such as lice and
mites, can reproduce directly on the animal. Other species, such as ticks, have
another reproduction strategy, in which
the juvenile forms live on the vegetation.
At certain stages, they need to encounter
an animal in order to complete their life
cycle and reproduce. Therefore, the control of external parasites is a combination
of measures directly related to the animals
with measures to reduce parasite levels in
the environment.

Make use of natural resistance of local breeds

Several local breeds have innate resistance against external parasites. See the corresponding section above, for internal parasites. Again, to reduce the impact of parasites, use local breeds instead of exotic animals or crossbreeds.

Support additional ethno-veterinary treatments and practices

Many local ethno-veterinary products and practices for parasite control exist in each region. These traditional remedies for the control of different types of parasites can be stimulated. Examples are:

* The use of chickens in the corral to rid the cows from ticks;
* Regular baths with ethno-vet herbal remedies.

Community control activities of external parasites

If an animal has only a few ticks, these can be carefully pulled off, making sure the mouthparts of the tick are removed. Rubbing ticks with a cloth soaked in kerosene or paraffin will make them drop off the host. Large numbers of ticks, mites and lice are best controlled by spraying or dipping the animal with a suitable treatment, based on medicinal plants or a chemical product. All of the flock or herd must be treated at the same time. In some cases, community external parasite control activities can be organized – especially of wool producing species – and sometimes in combination with other community gatherings.

6: Breeding and Selection

Breeding the best animals is a central challenge in any animal production system (Table 9.6). Farmers want healthy and high-producing animals that are adapted to their environment. In low-input systems, this challenge is especially great, because of the challenging environment with seasonal feeding shortages, specific parasites and diseases as well as the multiple roles of the animals. In this context, numerous local animal breeds have been developed over time.

The recommendations related to local breeding and selection in low-input and diversified livestock keeping are divided into two parts: (i) use and choice of breeds; and (ii) breeding management.

Use and choice of breeds

Local animal breeds are often despised as being unproductive and backward. In most

Table 9.6. Objectives and recommendations for improved breeding and selection in smallholder low-input and diversified livestock keeping.

Breeding	Objectives	Use of breeds	Breeding management
Low-input and diversified systems	Maintain important local breeds Make use of important traits of local breeds Effective selection Prevent inbreeding		
Recommendations for improvements		Breed selection on basis of local criteria Selection of better quality local breeds	Prevent inbreeding Timely castration Change males before mating with own offspring

countries that want to boost their animal production, a recipe of importing exotic breeds and crossbreeding programmes is almost standard procedure, even under smallholder conditions with severe limitations. This is done at the cost of the breeds that are used locally, which has resulted in severe loss of local breeds worldwide (Fig. 9.12).

One can say that especially in low-input animal keeping, the use of exotic breeds and crossbreeding is not the best option available. Experience has shown that – because of the severe limitations in terms of feeding and management inherent in this animal production system – improving local breeds through selection is a more viable and promising option. Bringing in exotic breeds and crossbreeding are options that imply a change from low-input into a more specialized animal keeping system. This will therefore be discussed in the following chapter (Chapter 10).

The myth of the 'unproductive' local breeds is a hard-to-crack notion, because indeed local breeds are usually less productive in conventional terms than the so-called exotic or high-yielding breeds. Some elements need to be taken into account in that respect, however (ANTHRA, 2008):

- Local breeds produce even under difficult circumstances and with very low inputs; therefore local breeds are especially productive in marginal areas.
- Local breeds do not only produce milk, eggs or meat – but also other essential products such as manure and draft power and transport.
- Local breeds are very diverse and often look more 'messy' than animals of standardized high yielding breeds (Fig. 9.13).
- Local breeds are usually smaller. This looks as if they are genetically degenerated.
- Because of deficient management, the mortality in offspring in local breeds is high; therefore, it seems as if they produce less.
- Inbreeding and deficient management is all too common in local breeds, which leads to problems, including low productivity.

Fig. 9.12. Raika pastoralists have been able to maintain several special breeds that are well adapted to their dryland conditions in Rajasthan in northern India. Credit: Ilse Koehler Rollefson.

Fig. 9.13. Because of high mortality, chicken of local breeds seem to produce only few chicks. In fact, these 'Naked Neck' local breeds are very fertile and productive, even under difficult circumstances. Credit: Ellen Geerlings.

- Local breeds often have special traits, which are more difficult to measure than productivity, such as drought resistance, disease resistance, high fertility and special product quality.
- The potential for improvement of local breeds through selection is high – but practical experience in supporting this process is still very limited.
- Any exotic high yielding breed was once a local breed!

The characteristics of local breeds are sometimes more appreciated outside the local environment. An example is the experience of local cattle breeds from India (Gir and Kankrej from Gujarat, and Ongole from Andhra Pradesh) that were imported into Brazil in the 1960s. Besides producing meat, these breeds were developed as excellent milk breeds after a process of selection. In fact, the world's best Gir cows today live in Brazil and give around 5500 l of milk on average per lactation. Comparing these with the neglected cousin in India, which does not yield more than 980 l, the Brazilian Gir yields roughly six times more (Sharma, 2010).

Genetic improvement of local breeds through selection

Improvement of animals of the local breeds can be very successful, with a careful look at the individuals that are available and using the local selection criteria of the desired traits in males and females. Over time, the desired characteristics can be enhanced within the local breeds, based on the roles of the breed and the specific challenges it needs to face. This process can be effectively supported. Animals of improved local breed, especially selected males, can then be re-introduced into their environment (Perezgrovas, 2006).

Breeding management

Prevent inbreeding

Many livestock keepers have effective knowledge and practice related to breeding. It is, however, common to find inbreeding in low-input husbandry. Related animals breed among themselves, gradually resulting in a genetic degeneration. It does not cause mortality directly, but can lead to deformed, weak and poorly growing animals that are more likely to die. Depending on the species, there are practices to deal with this problem, such as dividing animals into age groups. When this is not possible, it is necessary to take other measures such as:

- Timely castration of males, so they cannot mate with their mother/sisters;
- Regular exchange of reproductive males;
- Selection of breeding males and castration of all other male animals.

7: Protection and Housing

Efficient protection is another central element to reduce animal mortality effectively in low-input and diversified animal keeping. No major constructions are necessary – efficient constructions based on local materials have been developed by livestock keeping families (Table 9.7). A world can still be gained in this respect, however, as mortality related to predators, theft and trampling is excessively high in low-input livestock keeping. This is especially the case among young animals. Moreover, effective protection in simple constructions can reduce the risk of transmitting animal diseases to humans.

The recommendations related to protection in low-input livestock keeping are divided into two parts: (i) prevent predators, accidents and theft; and (ii) protection against weather conditions.

Protection against predators, accidents and theft

Protection during first weeks of life

Newly born and young animals are more affected by predators and accidents. Depending on the species, small investments can be made for the construction of temporary shelter and thus considerably reduce mortality. It is important to know the customs of the place and to see the innovations that some families have developed in this respect, each with advantages and disadvantages.

It can be observed that many women have experimented with a simple shelter for the chicks during their first three weeks of life. Their feed needs to be provided during

Table 9.7. Objectives and recommendations for improved protection and housing in smallholder low-input and diversified livestock keeping.

Protection	Objectives	Predators, accidents and theft prevention	Weather protection
Low-input and diversified systems	Reduced loss because of predators, theft and trampling Effective low-cost construction with local materials Prevent transmission of zoonotic parasites		
Recommendations for improvements		Protection of young animals during first weeks Protection during brooding and caring for young Night shelters Control between animal and man parasite transmission	Provide simple night shelters

this period, such as ground maize. Though this implies more work, the mortality of the chicks decreases significantly during this most dangerous period.

Protection during brooding

Adult animals also need special protection when they are brooding (Fig. 9.14) or caring for their young. Again, it is possible to build on special protection experience existing within the locality for good ideas and examples.

Protection against adverse weather

Shade and wind breaks in the field

Protection against sun, rain, wind, cold and lightning is important for reducing mortality, especially for young animals. Depending on the species, this can be provided by trees with wide shade areas, natural or tree wind breaks as well as the use of a roof or a little shed. Raised areas in times of flooding are necessary.

The construction and cleaning of these shelters are very important for avoiding parasites and other bugs, such as the assassin bug. Livestock on the open range needs protection from strong winds, cold winds, severe rain, flooding, and extreme heat and humidity to prevent lung disease especially.

Special night shelters

Night shelters are an essential element, especially for animals that roam around during the day, to protect them from theft, predators and adverse weather.

An example of special night shelter is the custom of penning animals, such as chickens, in the houses at night. Though this effectively protects the animals, epidemiological studies have shown that this custom also increases the incidence of diseases and parasites, such as assassin bugs and Chagas disease.

8: Special Care

Special care is another effective way to reduce animal mortality in low-input animal keeping. Special care is especially relevant around birth and for individual animals with a disease or other problem (Table 9.8). In this way, the foundation is laid for a healthy future and productive life. This only requires special knowledge and attention without the need for expensive constructions.

The recommendations related to special care in low-input livestock keeping are divided into two parts: (i) special care for sick animals; and (ii) special care before, during and after delivery.

Fig. 9.14. Chicken effectively protected from predators and weather during brooding. Credit: Ellen Geerlings.

Table 9.8. Objectives and recommendations for improved special care in smallholder low-input and diversified livestock keeping.

Special care	Objectives	Sick animals	Around delivery
Low-input and diversified systems	Increased survival rate of sick animals Reduced mortality of newborn Reduced disease female animal after birth Neo-natal bonding Reduced disease transmission around birth		
Recommendations for improvements		Separate sick from healthy animals Shade, water, fresh feed Ethno-vet/commercial treatments Disposal of dead animals	Observe carefully animals before birth Attend birth when necessary Check afterbirth Assure colostrum intake

Sick animals

Special care is very important to increase the survival rate of sick animals (see also Birmingham and Quesenberry, 2007).

Isolate sick animals

It is necessary to isolate sick and recovering animals from healthy stock and give them special attention. If mixed with healthy animals, there can be competition for feed and water, which slows down recovery.

Provide shade, fresh water and feed, treatment

It is essential to keep the sick animals in the shade and protected from wind and rain. The animals need to be supplied with fresh water all the time and some good quality food.

Disposal of dead animals

When an animal dies, it is important to dispose of the carcass in a correct way. This is especially relevant in the case of infectious disease, in order to reduce the chances of infecting other animals. In the case of death related to anthrax, for example – a highly contagious infectious disease of ruminants –

remnants of the animal can remain in the soil and water, and infect other animals, even years later. In this case, it is essential to prevent vultures from eating from the carcass, but rather bury or burn it.

Special care before, during and after delivery

Extra assistance around delivery

Extra care is especially relevant just before, during and after birth. Sometimes assistance is needed to ensure that delivery is normal and the young can breathe well. Moreover, the young need to be protected from cold and drink the first milk with special qualities. The feeding will also stimulate good bonding between mother and young. The umbilical cord needs to be cleaned and protected from infection. Finally, it is important to see that the afterbirth comes off normally.

Simple birthing pen

While many livestock young are born outdoors and under conditions of minimal care from the livestock owner, birthing under more controlled conditions can greatly reduce mortality. There is a balance that seasoned livestock owners know – that some mothers need to be left alone and not bothered, while others need help. A general rule of thumb, though, is to let nature take its course during birthing and let the mothers do what mothers are supposed to do – with a watchful eye from the caretaker.

All livestock giving birth need to be observed – whether on the range or in confinement – and assistance needs to be given if anything is not going smoothly with delivery.

- If the newborn is too large or in an unusual delivery position or if the mother is too small and too weak, assistance needs to be given right away.
- If the weather is bitterly cold or excessively wet, a dry spot out of the wind is a sanctuary.
- If predators or theft are common, safe areas can be provided.
- If the newborn are weak or if the weather is extremely cold, sheds with a means of warmth are necessary.

Facilitate feeding with first milk (colostrum)

The first feeding of milk needs to be assured within a few hours after birth. This milk with special antibodies – also known as colostrum – has special relevance in preventing disease in the first year of life. In order to have this quality, it needs to be consumed during the first few hours of life.

Special care for recently born piglets

Simple protective measures are also possible to reduce piglet mortality. The highest mortality of these animals is common in the first hours after birth, until 6 days after birth, and is in fact the major reason for piglet mortality in low-input systems. This occurs because of the cold, when the young look for heat and thus can be smothered by the mother when she wants to lie down. When a piglet is born, it can be temporarily placed separately and kept warm – for example under a warm light or using a bottle with hot water wrapped in a rag. After having delivered all the piglets, these can be placed back with the sow. During the first few days, the young need to be placed with the mother every 2 h to nurse. With small adaptations, such as placing a table where the young can sleep separately, the death of young piglets can effectively be avoided (van't Hooft, 2004).

References and Further Reading

ANTHRA (2008) Unpacking the poor productivity myth. Women resurrecting poultry biodiversity and livelihoods in Andhra Pradesh, India. Good Practice Brief SAGP25, South ASIA Pro-Poor Livestock Policy Programme, FAO/NDDB, Rome.

Bhandari, D.P. (2009) *Community Animal Health Worker Manual*. Heifer International, Little Rock, Arkansas.

Birmingham, M. and Quesenberry, P. (2007) *Where there is no Animal Doctor*. Christian Veterinary Mission, Seattle, Washington.

Flintan, F. and Cullis, A. (2010) Natural Resource Management Technical Working Group, Ethiopia. *Introductory Guidelines to Participatory Rangeland Management in Pastoral Areas*. Save the Children, US/FAO, Rome.

FRLHT and Tanuvas University (2010) Report on International Conference on Ethnoveterinary Practices. Mainstreaming Traditional Wisdom on Livestock Keeping and Herbal Medicine for Sustainable Rural Livelihood across Continents. Thanjavur, 4–6 January.

Gebru Tegegn, G. (2010) PARIMA reintroduces controlled burning, a traditional range management practice. In: Lammerink, M. and Otterloo-Butler, S. (2010) *Seeking Strength from Within – The Quest for a Methodology on Endogenous Development*. Compas network of ETC Foundation.

Hooft, K. van't (2004) *Gracias a los Animales Crianza Pecuaria familiar en America Latina, con casos de los Valles y el Altiplano de Bolivia (Thanks to the Animals – Family Level Livestock Keeping in Latin America with Case Studies from the Bolivian Valleys and Highlands)*. Plural Publishers, La Paz, Bolivia.

Karbo, N. (1999) Natural mineral licks to enhance livestock growth. CSIR Research Animal Research Institute, Tamale, Ghana. In: *Appropriate Technology*, volume 34, no. 1. Livestock in Development, Crewkerne, Somerset, UK.

Köhler-Rollefson, I. (2010) LPPS, Saving the camels of Rajasthan (www.pastoralpeoples.org).

Kyvsgaard, N., Luna, L. and Waagstein, T. (2001) Sostenibilidad de un programa de vacunacion contra la enfermedad de New Caste en Nicaragua. Report of 10th AITVM conference, August, Copenhagen.

Perezgrovas, R. (2006) Direct involvement of indigenous women in sheep improvement research in Chiapas, México. Prize winning poster in German Tropentag 2006, Bonn: Prosperity and Poverty in a Globalized World – Challenges for Agricultural Research.

Sharma, D. (2010) Holy cow, Acclaimed abroad, despised at home (part 1 and 2) Blogpost on Arise India Forum. www.ariseindiaforum.org/cow_protection.php

WHO (2005) The Control of Neglected Zoonotic Diseases – A route to poverty alleviation. Report of a Joint WHO/DFID-AHP Meeting. WHO, Geneva.

10

Recommendations for Optimizing Smallholder More Specialized Systems

Learning Objectives: Understanding
- The need to adapt recommendations to livestock-keeping strategy
- The main causes of animal mortality and low productivity in more specialized livestock keeping
- Recommendations for stimulating more specialized livestock-keeping systems in eight special areas

Adapt Recommendations to Local Circumstances

It is important to analyse the type of animal husbandry used within a family or community, as well as the local circumstances at hand, before embarking on activities in support of the animal husbandry practices. Because of the differences in objectives between systems, as explained in the previous chapters, it is necessary to adapt the recommendations similarly. This will be done for more specialized animal keeping systems in this chapter. The improvements of low-input and diversified systems are discussed in Chapter 9. The aspects related to marketing will be detailed in Chapter 11.

As indicated in the previous chapters, most livestock-dependent limited resource people can be found in the low-input and diversified smallholder farming and pastoralist systems. In addition, the same family may be employing a more specialized type of animal production of one selected species. This more specialized system will have a lower input level than large-scale commercial farms or ranches, but the logic of this livestock-keeping system – in terms of productivity focus – coincides more with the high-input commercial livestock keeping than with low-input and diversified farming.

Main Goal: Increased Animal Productivity and Farm Efficiency

In more specialized production systems, the intensity of inputs, labour and other costs are increased when compared with low-input and diversified livestock-keeping systems. Production per animal is normally higher and mortality is generally lower than in low-input and diversified livestock keeping. The more specialized smallholder livestock-keeping systems can be improved with the goal of increasing animal production and farm efficiency within the integrated agricultural system (van't Hooft, 2004).

In Box 10.1, the main reasons for mortality in low-input and diversified and more

Box 10.1. Main causes of mortality and loss in low-input and more specialized livestock-keeping systems.

Low-input systems	More specialized systems
Nutritional deficiencies, especially during dry periods	Infectious diseases
Lack of pasture	Acute conditions new to the farmer, such as milk fever
Water deficiencies	Birthing difficulties because of offspring with big size
Infectious diseases	Acute udder infections (mastitis)
Internal and external parasites	Culling because of infertility, lameness or injury
Breeding deficiencies	
Lack of protection	
Lack of care in special moments (birth, illness)	

specialized systems are shown. Because of different management of the animals, they show some remarkable differences.

The ways to support more specialized systems are shown in Fig. 10.1 and are based on the recommendations to reduce mortality for supporting low-input and diversified systems, as mentioned in Chapter 9. These are the minimal requirements. This chapter explains the extra recommendations for more specialized livestock-keeping systems. Please note that this is an overview of possibilities of improved management practices rather than a complete guide.

1: Improved Animal Nutrition

Upon making the change from low-input and diversified to more specialized animal husbandry, the first thing to change, generally, is animal nutrition; both the quantity and quality of feed needs to be improved to maintain an adequate level of nutrition throughout the year. Dry and wet seasons make less difference in nutritional support than in low-input systems. The aim is to provide an optimally balanced feed ration from commonly available nutrient stocks, including grains, silage, hay and supplements. The ingredients are either grown on the farm or purchased through area markets.

Producers all know that it is very difficult to provide completely adequate nourishment throughout the year, especially for families that have recently changed from diversified husbandry to more specialized husbandry. To assure good management of

the animals requires continuous investments of capital and labour, which is not always easy to guarantee. Moreover, at the beginning of the process of change from one system to the other, families are insecure about their debts and how to access new markets. At this period these families need extra support.

The recommendations for animal nutrition in more specialized livestock keeping are divided into two parts: (i) dry season nutrition; and (ii) mineral supply.

Dry season nutrition

Good-quality animal nutrition year round may well be the most challenging factor for more specialized livestock keeping for all producers. For low-input systems, the nutritional objective during the dry season is to reduce mortality and weight loss. This is done by taking advantage of available fodder stocks, such as straw from grain crops, leguminous trees and other agricultural by-products.

These strategies that are described in detail in Chapter 9 can be maintained and improved in more specialized systems:

• Improved straw storage and feeding;
• Supporting local feeding innovations and traditions;
• Improved use of kitchen leftovers;
• Green forage, such as oats, barley and lucerne;
• Hay making;
• Cheap and easy to obtain by-products, such as wheat bran;
• Use of feeder troughs.

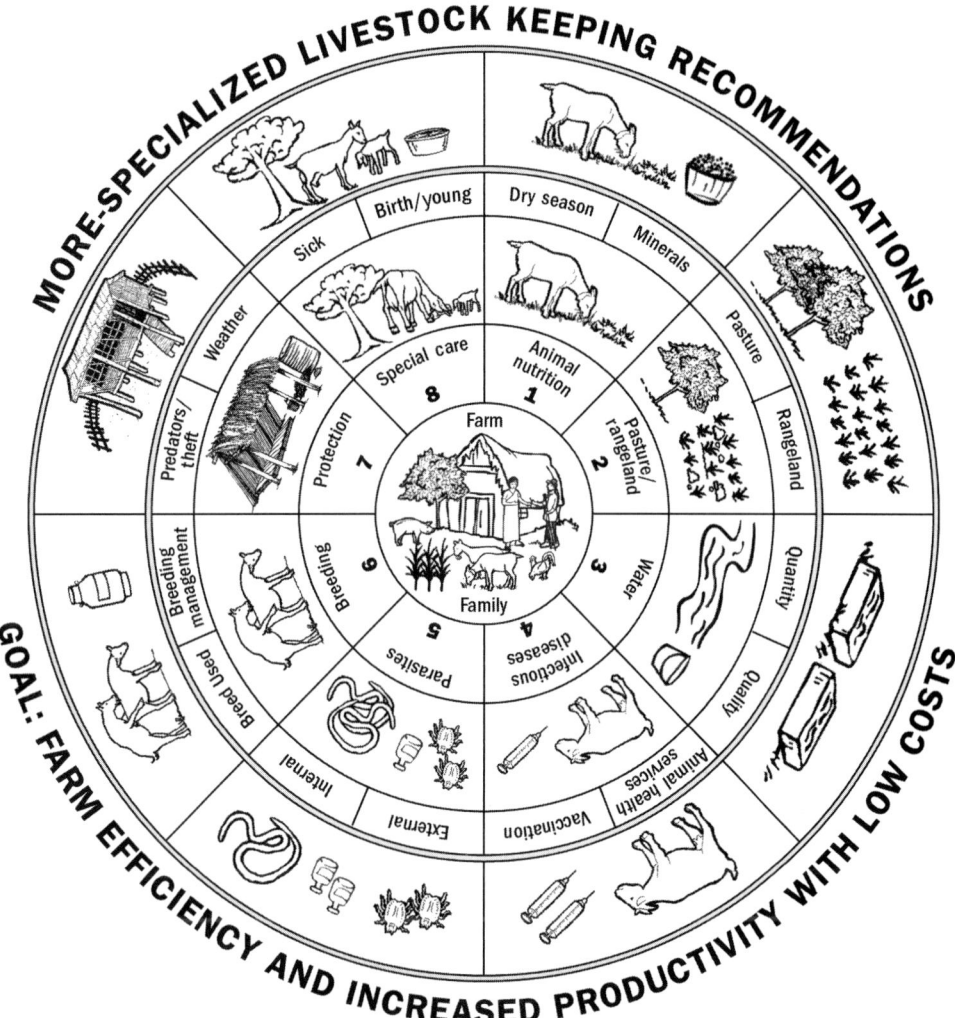

Fig. 10.1. The wheel of animal well-being and production for more specialized livestock keeping, showing the eight major areas of improved management in more specialized animal keeping. They build on the minimum recommendations for low-input livestock keeping.

Additional feeding strategies in more specialized systems are described below.

Lucerne with irrigation

Lucerne is a high-quality legume that can be used to feed the animals throughout the year. Different varieties of lucerne are used depending on the climatic and soil conditions in each area. Lucerne requires irrigation during the dry months, and is therefore a crop that requires relatively high input.

Because of its quality, it is a well accepted crop in many regions of the world. It is mostly used in cut-and-carry systems. In some cases, lucerne is used in controlled grazing, though the crop does not resist heavy grazing.

Elephant or Napier grass

Fodder crops are often planted in the case of initiating a zero-grazing system. In this case (Fig. 10.2), the Napier grass is best planted as

Table 10.1. Comparing objectives and recommendations for improved animal nutrition in smallholder low-input and diversified livestock keeping (with more specialized animal husbandry).

Animal nutrition	Objectives	Dry season nutrition	Mineral supply
Low-input and diversified systems	Reduced mortality in dry season Reduced weight loss Increased resistance to drought		
Recommendations for improvement		Agricultural by-products storage and feeding Support local feeding innovations Plant leguminous trees Improved use of kitchen leftovers Green forage Hay making Feeder troughs	Provide ordinary salt Home-made mineral blocks Vitamins
More specialized systems	Better nutritional status year round Improved reproduction rate Special feeding of young stock		
Recommendations for improvement		Local production of balanced feed Improved straw feeding Hay/silage making	Complete mineral supplements Vitamins

Fig. 10.2. Feeding with elephant grass in a zero-grazing unit.

close to the zero-grazing facility as possible. This reduces labour in transporting the materials.

If the fodder grass is planted in rows, a shallow trench in between each row can facilitate watering. Manure from the pens can also be used in these trenches to fertilize the grass.

A way to judge the amount of land needed for one cow is the following. About 0.4 ha (1 acre) of Napier grass is enough for each local dairy cow per year in most areas. If the Napier grass is intercropped with *Desmodium* or other legumes, 0.3 ha (0.75 acre) is enough for one cow; 0.4 ha of Napier and *Desmodium* is enough to feed a cow and a calf. Other local grasses can be evaluated as substitutes for Napier and *Desmodium* (Bhandari, 2009).

Silage

Silage is the storage of green feed in a pit covered with plastic, so that it ferments and, as such, maintains its quality. The most common silage is of maize, but other forages can be stored. Good feed must contain dry stubble to ensure fibre content. Silage is utilized in more specialized dairies because it involves greater expense: digging the pit, paying the farmhands, the chopper and the plastic.

Cottonseed

Cottonseed is a by-product of cotton production and can be purchased relatively cheaply in cotton-producing areas at harvest time. It is a high-energy feed and, together with bran and mineral salt, provides an almost completely balanced diet for dairy cattle. Before giving this feed, straw or maize husks must be fed to the cattle to balance their diet with fibre. Purchasing, transporting and storing the seed require a large investment, which is often done in cooperative groups or specialized dairies. Feeding cottonseeds can cause problems, especially in young animals, because of the insecticides utilized in the cultivation of cotton. The effects on humans are unknown.

Urea

Dry fodder crops can be improved in their feeding quality by adding urea and molasses (sugarcane syrup – a by-product of the sugarcane industry) to increase palatability and nitrogen content for cattle. This practice is not without risk, as urea is toxic when applied in too large quantities. Animals need to be accustomed to this product little by little.

Making balanced feed

It is possible to optimize diets in more specialized husbandry systems, seeking a balance between maximum production and cost optimization using relatively inexpensive and easily accessible feed.

Balanced rations can also be made from a basis of maize or sorghum, defatted soybean flour, defatted cottonseed flour or sunflower and mineral salt. Its quality can vary widely depending on its content. Generally, it is of excellent quality, but at a high price. An advantage is that it can be purchased in small quantities.

Here is an example of a complete yet simple dairy cow feed formulation containing all of the nutrients needed for optimum maintenance and production, using locally available feed ingredients (Bhandari, 2009):

* 2 parts of rice bran or wheat bran;
* 1 part of maize, millet or sorghum;
* 1 part of mustard/cottonseed/sunflower cake or any bean. Grind all feed ingredients together well and feed at the rate of 1–2% of the body weight.

Mineral supply

The strategies for low-input livestock keeping that are described in detail in Chapter 9, can be maintained and improved in more specialized systems:

* Provide ordinary (kitchen) salt;
* Preparation of simple mineral block;
* Vitamins.

Additional mineral and vitamin supplement strategies in more specialized systems are shown below.

Provide mineral salt

In more specialized systems, it is necessary to supplement feeds not only with kitchen salt but also with other minerals to prevent weight loss, maintain high milk production, growth rates and reproductive functions. It is necessary to maintain complete mineral supplements year round, for example through commercial or homemade mineral blocks.

Reducing the amount of mineral salt in the feed of dairy cattle for a few months is a common practice to reduce costs in low-input dairy husbandry. Nevertheless, under the logic of more specialized husbandry, it is more favourable to sell down livestock numbers in times of stress in order to maintain the level of nourishment for the remainder of the herd. The lack of mineral salts in dairy cattle, especially higher-producing breeds, generates problems with reproduction, yielding difficulties at calving, smaller birth weights or stillbirths, which constitute a loss many times more valuable than the saved mineral salt.

Mineral supply is essential for all kinds of animal keeping, and even more so in more-specialized livestock production (Fig. 10.3). The lack of mineral salts in dairy cattle, for example, especially higher-producing breeds, generates problems with

diseases resistance, parasites as well as reproduction. Though less tangible, this constitutes a loss many times the value of the cost of the mineral salt.

Much needs to be understood about the need for different minerals in different species of animals and at different stages of production and reproduction. For example, sheep cannot tolerate as much copper in their feed as do goats, so using the same copper supplemented mineral mix for both species requires an educated look at the ingredients of the product. Moreover, mineral deficiencies also depend on the mineral contents of the soil and forage of the particular area. For example, if an area is particularly deficient in selenium, then reproductive functions and some illness can result. Species differences are also common.

Provide vitamins

In more specialized livestock keeping, the separate provision of vitamins is less necessary than in low-input keeping, especially when animals are provided with balanced feeds and other high-quality products.

2: Improved Pasture and Rangeland Management

Upon making the change from low-input or more diversified husbandry to more specialized husbandry, improved pasture

Fig. 10.3. Locally made salt block in the stable of a smallholder dairy farm.

and rangeland management is another essential element. The objective of more specialized systems is a straightforward, year-round (or, at best, season long) availability of good-quality pasture grasses, forbs and brush for all animals (Table 10.2). This needs to be produced with good mineral efficiency, and be based on healthy soils, with good soil life.

The recommendations for pasture and rangeland management in more specialized livestock keeping are divided into two parts: (i) pasture management; and (ii) rangeland management. Please note that is not intended to be a complete guide for this extensive topic, which is also under continuous discussion.

Improved pasture management

Good pasture management is a challenging factor for more specialized livestock keeping. In low-input livestock keeping, pasture management is not much used. The only common form of pasture management in low-input systems is controlled through grazing, for example, by tying the animals to a pin and regularly moving them. Grazing young stock that are already weaned away from the older stock helps to cut down on the transfer of internal parasites. These strategies that are described in detail in Chapter 9 can be maintained and improved in more specialized systems:

Table 10.2. Comparing objectives and recommendations for improved pasture and rangeland management in smallholder low-input and diversified livestock keeping (with more specialized animal husbandry).

Pasture and rangeland management	Objectives	Pasture	Rangeland
Low-input and diversified systems	Reduced overgrazing and soil erosion Reduced bush encroachment Increased carrying capacity Increased resistance against drought Community organization		
Recommendations for improvement		Controlled grazing Zero-grazing system	Reviving communal grazing control Fencing off grazing areas Rotational grazing Special grazing areas for dry period Controlled and pre-scribed burning
More specialized systems	Sufficient fodder available year round Good-quality fodder Good N and P efficiency Increase soil fertility and soil life		
Recommendations for improvement		Plant fodder crops Pasture rotation Special pastures for young stock Zero-grazing system Efficient fertilization of pastures	Effective weed control

- Controlled grazing;
- Zero grazing system.

Additional pasture management strategies in more specialized systems are described below.

Weed control

Weed control is often taken care of by strategic rotation of animal species or mechanical means. Only in severe infestation cases are chemical means used.

Fertilization of pastures

Fertilization of grazing lands varies in more specialized systems and is usually limited to the use of livestock manure, either directly applied by grazing animals or spread mechanically from composted materials. This provides readily accessible organic materials back to the soil for decomposition.

Commercial fertilizer granules composed of nitrogen, phosphorus and potassium (NPK) along with lime to adjust soil pH can also be applied to rangelands. These should be preceded by an assessment of the soil needs by a professional soil scientist or sample analysed in a laboratory.

Soil humus health must also be considered. Commercial application of a specific NPK or lime fertilizer lacks this component. Organic material sources provide improvements to soil tilth, or suitability to be tilled, that helps to hold soil moisture and build a body to the soil while the basic chemicals that describe soil fertility can come from either source.

Milk and carry system

For dairy animals, pasture management systems can be designed so that cows are gathered for milking at designated places in the pasture. Milking equipment can be kept in the pasture, where animals are either tethered for milking or milked freely. Instead of a *cut and carry* system (see also zero-grazing), this is a *milk and carry* type of management, which is also common in low-input livestock keeping.

Zero-grazing system

In order to reduce the pressure on the pastures, sometimes it can be advantageous to use the zero-grazing system in which the feed is brought to the animals. This implies a transition to a more specialized animal keeping system. Zero-grazing systems can be used with any of the grazing species, such as cattle, goats and sheep, and even with chickens, pigs, rabbits and other animals kept for production and marketing. Animals are housed in an appropriately sized and simple shed with a slatted wood or hard dirt floor all made from local materials. Animal droppings fall through the slats onto the ground below and can be collected easily for composting or direct fertilization of crops.

(See also later section on Protection and Housing.)

Improved rangeland management

In low-input systems, rangeland management objectives relate to reducing the encroachment of brush, reducing soil erosion, increasing the carrying capacity of the land and managing through drought conditions.

The strategies for low-input and diversified livestock keeping that are described in detail in Chapter 9 can be maintained and improved in more specialized systems:

- Reviving communal grazing control;
- Use of traditional animal species, such as camelids;
- Fencing off grazing areas;
- Rotational grazing;
- Reviving indigenous practices to reduce bush encroachment, such as prescribed fire.

Additional recommendations for more specialized animal keeping are given below.

Mob grazing

Pasture rotation systems are many and varied. Sometimes called *mob grazing*, the aggressive control of where ruminants graze can be quite beneficial to the rangeland and can be a positive benefit to reduction of greenhouse gases while improving soil carbon retention. Hooves loosen the soil as the animals graze together (Fig. 10.4). The heavy deposit of manure and urine in the mob-grazed area leaves the ground ready for rains and rapid regrowth of vegetation, once the animals are moved to their new grazing spot.

The feature of mob grazing that must be kept in mind is that the animals must graze down the vegetation within a designated area thoroughly before being moved. They cannot be allowed to eat only what tastes good, but all vegetation. Sometimes this is facilitated by concurrent or sequential grazing of cattle, goats and sheep. These three species all prefer different types of forage and can be managed so that all grasses, forbs and shrubs in a given area can be fully utilized.

When well managed, mob-grazing systems actually allow greater utilization of pastures so that more animal growth results from a given area than when animals are given free choice to the entire area at one time. In this type of management system one must be aware of the specific optimum grazing depth so that plants are not eaten down to the point that they will not grow back quickly or grow back at all. Each plant has its optimum height of vegetative material above ground that is needed for the root to stimulate regrowth. Too much grazing down will kill or severely retard regrowth.

3: Water Provision

Dirty water or insufficient water is a factor that seriously limits any form of animal husbandry. This essential and basic element is too often neglected and under-estimated, especially in smallholder conditions (Table 10.3). The possibilities for improvement logically depend on the conditions of each place. Bad water quality or lack of water can greatly reduce productivity and can go relatively unperceived until too late. In more specialized systems, extra attention needs to be given to water quality and quantity.

Fig. 10.4. Pasture rotation systems require extra inputs but can be beneficial for the rapid regrowth of the vegetation. Credit: Ann Wells.

Table 10.3. Comparing objectives and recommendations for improved water provision in smallholder low-input and diversified livestock keeping (with more specialized animal husbandry).

Water	Objectives	Access to water	Water quality
Low-input and diversified systems	Regular water uptake Water quality sufficient Pollution of human water sources prevented		
Recommendations for improvement		Water 1–2 times a day	Prevent polluted drinking water for animals
		Opt for animal species that require little water	Prevent pollution of water for human use by animals
More specialized systems	Good water availability Quality year round		
Recommendations for improvement		Continuous access or provide 3–4 times per day	Prevent pollution of drinking water by crop chemicals and artificial fertilizers

The recommendations for water provision in low-input livestock keeping are divided into two parts: (i) access to water; and (ii) water quality.

Access to water

The strategies for securing access to water in low-input livestock keeping are described in detail in Chapter 9, and can be maintained and improved in more specialized systems:

- Regular daily watering;
- Opt for animal species that require less water.

Additional recommendations for more specialized animal keeping are described below.

Access to water at least three to four times a day

In more specialized systems, the need for water is larger than in low-input livestock keeping, because of higher levels of production. The animals need water refreshment at least three or four times a day, with more frequent access in extremely hot weather.

Water – how much?

A good rule of thumb is to provide 1/20 of the animal's body weight in the weight of water each day (Fig. 10.5). One can figure out rough estimates for each animal using this conversion: 1 l of water weights about 2 pounds; 1 gallon of water weighs about 8.35 pounds. Thus, an 800-pound (365-kg) cow needs about 5 gallons (or 18 l) of clean water per day.

Water quality

The strategies for securing water quality in low-input livestock keeping are described in detail in Chapter 9, and can be maintained and improved in more specialized systems:

- Prevent polluted drinking water for animals;
- Prevent pollution of human drinking water sources by animals.

Water sources need to be protected from all sources of contamination, whether from barnyard waste materials or other pollutants. This is easier said than done and depends on local circumstances. Extra efforts in this

Fig. 10.5. Regular water supply is a major requirement for all animals, but especially so in more specialized animal keeping. High-producing animals are less resistant to water shortages than low-producing ones.

direction are, however, efficient and low-cost ways of improving livestock production and reducing animal mortality.

4: Control of Infectious Diseases

Contagious infectious diseases with high mortality are a common problem in low-input livestock keeping as well as in more specialized livestock keeping. Each animal species has one or two major infectious diseases that are often possible to prevent relatively easy through regular vaccinations. In order to accomplish the aim of reducing animal mortality, it is necessary to focus especially on the control of these infectious diseases (Table 10.4).

The objective of infectious disease control in more specialized systems is to reduce

the incidence of various diseases, besides these most common contagious diseases, without overlooking zoonotic infections (diseases transmitted between animals and man). Differences between the low-input and more specialized systems begin to appear in the practices of prevention and treatment.

The recommendations for the control of infectious diseases in more specialized livestock keeping are divided into two parts: (i) animal health services; and (ii) vaccination.

Animal health services

The strategies for securing animal health services in low-input livestock keeping are described in detail in Chapter 9, and can be maintained and improved in more specialized systems:

- Supporting ethno-veterinary practices and practitioners;
- Promote synergy between traditional and modern remedies;
- Train community-based animal health workers;
- Increase awareness about zoonosis.

Additional recommendations for more specialized animal keeping are described below.

Ethno-veterinary practices in more specialized livestock keeping

The collection and use of local healing knowledge and remedies is the practice of ethno-veterinary medicine. Ethno-veterinary animal health care – like commercial animal health care – can be broken into three basic areas of importance:

- The prevention of disease, which involves most of the practices of livestock management, from housing, nutrition, husbandry, environmental management as well as preventive vaccines;
- Disease treatment;
- Animal health surveillance. This is the identification of diseases and conditions with the aim of preventing spread, controlling infections and public health safety.

Table 10.4. Comparing objectives and recommendations for improved control of infectious diseases and animal nutrition in smallholder low-input and diversified livestock keeping (with more specialized animal husbandry).

Infectious diseases	Objectives	Animal health services	Vaccination
Low-input and diversified systems	Reduced incidence of zoonosis Reduced animal mortality because of infectious disease Promote synergy between traditional and modern remedies Improved access to local animal health services		
Recommendations for improvement		Support ethno-vet service Train community animal health workers Awareness of zoonoses	Vaccination of one or two major diseases
More specialized systems	Improved use of ethno-vet medicine Improved use of commercial medicine One Health – human and animal medicine join forces		
Recommendations for improvement		Ethno-vet practices strengthened Training in improved use of commercial medicines Disease surveillance, Monitoring and recording of disease incidence	Extended vaccination programmes

Ethno-veterinary practices and community healers are most common in the low-input systems, while commercial preparations and professionally trained practitioners tend to be more common in the more specialized systems – especially with high value animals (Fig. 10.6). At the same time, in all livestock-keeping systems, even specialized high-input systems, farmers' local knowledge and experience is a valuable asset when it comes to animal health practices.

Community-based animal health workers in more specialized livestock keeping

Adequately trained community members can provide effective service to community livestock keepers, including those with more specialized livestock keeping, for a small fee. Community animal health workers can make a lifelong profession out of this work that supports their family and supports the community. In some cases, they set up small shops in their communities with veterinary and agriculture supplies and equipment (Birmingham and Quesenberry, 2007).

Training, retraining and monitoring of community animal health workers is managed by NGOs, farmers' organizations or government extension service.

Often the community animal health workers combine traditional and modern medicine, but this is not always the case. They will especially favour modern medicine when this is their major source of income. Sometimes farmers with very high

Fig. 10.6. The knowledge on and use of medicinal plants is an important resource in most smallholder livestock-keeping systems, including the more specialized ones. Credit: M.B.N.Nair.

value animals will prefer a formal veterinarian over a community animal health worker.

Conventional professional animal health care services

As more high-input and specialized livestock-keeping units start emerging in a region, the call and financial opportunities for formally trained veterinary practitioners begin to grow. These are often linked to commercial association of veterinary products, and tend to focus on modern and large-scale specialized livestock keeping, rather than low-input and diversified livestock keeping.

These conventionally trained veterinary caregivers can play an essential role in working together with local healers, community animal health workers and farmers in facing the following challenges:

- Training of improved use of commercial medicine;
- Monitoring and recording of disease incidence;

- Strengthening ethno-vet practices by appropriate combination with conventional medicine.

Improved used of commercial medicines

The indiscriminate and uncontrolled use of commercial products in smallholder livestock keeping is a major problem that in time can lead to human health problems related to residues in milk, meat and eggs. This is often complicated by the quality of the products being in question. If there is no strong regulatory system in the country, counterfeit and low-price medications may be abundant. Good training and mindful compliance is a safeguard against drug misuse (Fig. 10.7).

Common problems with poorly regulated commercial products are:

- *Low quality of products sold*: Not all manufacturers follow the required standards in the manufacture of products.
- *Low quality because of poor storage*: Medicines and vaccines should be

Fig. 10.7. The indiscriminate and uncontrolled use of commercial products, such as antibiotics and hormones, in smallholder livestock keeping is a major problem. In time, this can lead to serious human health problems related to residues in milk, meat and eggs.

stored in a clean, safe, temperature-controlled environment. If refrigeration is required to maintain potency, then this must be observed at all times; this is especially true for most vaccines.

- *Expiration dates*: Commercial products from legitimate pharmaceutical and biological (vaccines) manufacturers carry an expiration date, beyond which the potency and safety of the product cannot be supported by the manufacturer. Products should be destroyed when they have reached their expiration date.

- *Inadequate use*: Commercial medicine needs to be used according to the statements on the label. This may not happen, in some cases. Because of their cost, antibiotics, for example, are often not applied the number of days indicated, but only during the period with visible symptoms. This will, however, increase the incidence of resistant microbes to this antibiotic. The same is true for anti-parasitic treatments. Poor compliance in following label recommendations is one of the most critical problems related to the use of commercial medicine in conditions of low-input or more specialized livestock keeping.

- *Indirect hazards to human health*: Follow milk withholding and slaughter withdrawal times listed on the label. This is the responsibility of the owner and other health professionals. If milk withholding times and slaughter withdrawal times are not observed after the treatment, the meat and milk consumed can then contain residues of the medication, which can cause various health risks to the consumer. There are already major threats because of multi-resistant strains of microbes (Kumarasamy *et al.*, 2010).

- *Direct hazards to human health*: Several products, especially insecticides used for external parasite control, are hazardous to people or to the environment if they are not mixed or used according to the label. When using these types of product, protective clothing, gloves or eyewear are advised.

One Health concept

It is becoming much more evident that human health, animal health, public health and environmental health are all connected. This recognition has led to the emergence of an initiative that is variously known as One Health or One Medicine and is bringing practitioners of each field together.

One Health may seem far removed from the low-input livestock systems, however it has immense implications in villages with limited resources, especially related to the control of zoonotic diseases (WHO, 2005). When veterinary and human health services join forces, for example vaccinations against rabies can be more efficient if they cover dogs and cattle in one community at the same time (Fig. 10.8).

The One Health concept is a worldwide strategy for expanding interdisciplinary collaborations and communications in all aspects of health care for humans, animals and the environment. The synergism achieved will advance health care for the 21st century and beyond by accelerating biomedical research discoveries, enhancing public health efficacy, expeditiously expanding the scientific knowledge base, and improving medical education and clinical

Fig. 10.8. Rabies is one of the most serious common zoonoses still very prevalent in many developing countries that can be passed on through the bite of an infected dog. Joint action of medical and veterinary services is required for successful control. Credit: Florencio Pérez del Barco.

care. When properly implemented, it will help protect and save untold millions of lives in our present and future generations (Be.Troplive, 2010).

Disease prevention through vaccination

In more specialized animal keeping, control of infectious diseases through vaccination is extra-important, because of the high individual value of each animal and the larger number of animals kept together. The strategies for control of infectious diseases in low-input livestock keeping are described in detail in Chapter 9, and can be maintained and improved in more specialized systems, for example for swine fever in the case of pigs; Newcastle disease in the case of chickens; and blackleg, hemorrhagic septicaemia and anthrax in the case of cattle.

Additional vaccination against major infectious diseases

In more specialized systems, the risk of infectious disease is higher because of higher animal numbers. Especially when exotic breeds are introduced, there is a need for vaccinating against other infectious diseases. The more specialized the system, the larger the number of diseases against which the animals are vaccinated. The vaccines used are often against diseases that do not only cause direct mortality but also affect the animals in their productivity. According to animal species, the vaccination schemes need to be strictly applied.

5: Parasite Control

Parasites – both internal and external – are a common problem in low-input livestock keeping as well as in more specialized livestock keeping (Table 10.5). In general, parasites have a more devastating effect on young animals and exotic breeds – such as Holstein cows or Yorkshire pigs – than on adults and animals of local breeds. They also affect weak and malnourished animals

Table 10.5. Comparing objectives and recommendations for improved parasite control in smallholder low-input and diversified livestock keeping (with more specialized animal husbandry).

Parasite control	Objectives	Internal parasites	External parasites
Low-input and diversified systems	Reduced incidence of internal and external parasites Prevention of parasitic zoonoses Reduced loss of young stock Improved leather quality		
Recommendations for improvement		Make use of natural resistance of local breeds Reducing parasite incidence in grazing and feeding areas Parasite control especially in young stock Support ethno-vet remedies for parasite control	Make use of natural resistance of local breeds Use of medicinal plants for parasite control (ethno-vet) Community control activities (dips)
More specialized systems	Low incidence of internal and external parasites Special care for young stock Prevent resistance against commercial medications		
Recommendations for improvement		Regular treatment all stock Medicinal plants and commercial medications	Regular dips/spray of all stock Medicinal plants and commercial medications

more than healthy ones. Moreover, in warm and humid climates, parasites are more prevalent than in dry and cold climates. Therefore, incidence and the necessary control measures vary according to zone, species, breed, age and general state of the animals. For this reason, specific guidelines are difficult so consult local livestock holders, extension personnel and local healers.

Another problem related to both internal and external parasites in animals is that some of them can pass to people (parasitic zoonoses).

In low-input systems, the objective for control is usually the reduction of incidence to protect young stock and the quality of products and by-products, such as meat and hides. Methods of treatment are often herbal remedies or the reliance on natural resistance in local breeds for both internal and external parasites.

In more specialized systems, a more aggressive approach to parasite control needs to be taken. Even low incidence of external and internal parasites can draw down the resistance and growth of animals leading to loss of profit and death of the animals. Animals from highly productive breeds are especially susceptible. As in all systems, young growing stock usually suffers the most from parasite infestations.

The objective of all control mechanisms is to limit the effect of parasites to the extent that they do not affect growth and production. In the highly productive breeds, this usually requires regular anti-parasite treatments, which is becoming less effective because of resistance being developed by the parasites to the medical compounds used. Much can be done by natural means, such as strategic control of the grazing areas of the animals, use of novel natural substances and culling of individuals that carry the heavier loads of parasites.

The recommendations for parasite control in more specialized livestock keeping are divided into two parts: (i) control of internal parasites; and (ii) control of external parasites.

Control of internal parasites

Most types of internal parasites reproduce through a life cycle including eggs, larvae, pupae and adults. Many of the life cycles take place by deposition of manure on grassland followed by consumption of pre-infective or infective stages that further mature back in the animal's body. Eggs can survive outside the body for a certain period, and develop through several stages and life forms. Finally, the juvenile forms climb up the plants to be eaten by livestock. This process is greatly enhanced by high temperatures and humidity. As a result, the control of parasites in all animal keeping systems is a combination between animal-related measures and the control of grazing.

The strategies for controlling internal parasites in low-input livestock keeping are described in detail in Chapter 9, and can be maintained and improved in more specialized systems:

- Make use of natural resistance of local breeds;
- Reduce parasite incidence in grazing and feeding areas;
- Special focus on young animals;
- Support ethno-veterinary treatments and practices for parasite control.

Additional recommendations for more specialized animal keeping are described below.

Regular treatments, especially in young stock

Internal parasites have a dramatic effect on young animals, of both exotic and local breeds – though the effect on young animals of exotic breeds is more dramatic. The young become thin with swollen bellies. They grow abnormally slowly and they may even die. Parasitism, in combination with insufficient nutrition and the use of exotic breeds, is the major reason why livestock-keeping projects fail, especially in marginalized areas. Young animals in all animal production systems (Fig. 10.9) need extra treatments and care to reduce parasite numbers – while good nutrition can largely help to overcome the effects of parasites.

Timing of treatments

Timing of any anti-parasite treatment is critical for effective control. Thus, learning the life cycle stages is a key factor in use of internal parasite preparations. Treatment is often done prior to turning animals out for grazing in order to diminish infestation of the fields with parasite eggs, and subsequent exposure of the animals.

Treat new animals coming into the flock

All new animals need to be treated before joining an existing flock, in order to prevent introduction of new parasites.

Combine treatment with pasture rotations and good-quality feeding

Regular rotation of pastures, in combination with regular treatments and good-quality feeding, are essential elements in internal parasite control, especially in young animals of crossbreeds or exotic stock.

Prevent resistance against commercial products

Improper and too frequent use of commercial anti-parasite products can lead to the development of resistance of the parasites against

Fig. 10.9. Young animals in all animal production systems – and especially those of exotic breeds – need extra treatments and care to reduce parasite numbers – while good nutrition can largely help to overcome the effects of parasites. Credit: Ellen Geerlings.

the chemical composition of the drug. This can lead to multi-resistant parasites – which are extremely difficult to control.

External parasites

External parasites, such as lice and mites, can reproduce directly on the animal. Other species, such as ticks, have another reproduction strategy, in which the juvenile forms live on vegetation. At certain stages, they need to encounter an animal in order to complete their life cycle and reproduce. Therefore, the control of external parasites is a combination of measures directly related to the animals with measures to reduce parasite levels in the environment.

When deciding on treatment regimens, it is important to treat all animals in a group. Treatment of one or a few animals in a group is usually short lived. Truly effective and long-lasting internal parasite control will take a combination of using animal breeds with some resistance against existing parasites, effective bush control, pasture rotations, prevention of co-mingling, strategic use of systemic commercial dips or sprays, and ethno-veterinary antiparasitic preparations.

The strategies for controlling internal parasites in low-input livestock keeping are described in detail in Chapter 9, and can be maintained and improved in more specialized systems:

- Make use of natural resistance of local breeds;
- Support ethno-veterinary treatments and practices for parasite control;
- Community control activities of external parasites.

Additional recommendations for more specialized animal keeping are given below.

Treat new animals coming into the flock

All new animals need to be treated effectively before joining an existing flock, in order to prevent introduction of new parasites. Avoidance of co-mingling animals in markets or communal pastures is also a wise method to prevent transmission.

Combine treatment with pasture rotations, bush control and good-quality feeding

Regular rotation of pastures, in combination with regular treatments and good-quality feeding, are essential elements in internal parasite control, especially in young animals of crossbreeds or exotic stock. Bush control, especially controlled burning of old vegetation, can be another way of reducing tick numbers.

Past recommendations for more specialized livestock system parasite control often included frequent spraying or dipping of animals with certain commercial parasite control medications. This has led to the problem of some flies, ticks, lice and mites becoming resistant to the preparations that were regularly used. Since those products are no longer fully effective, such multi-resistant external parasites have become major problems (Fig. 10.10). Regular treatments are no longer a recommendation for parasite control products, even with novel rotation schemes of parasite product.

6: Improved Breeding and Selection

Breeding the best animals is a central challenge in any animal production system. Farmers want healthy and high-producing animals that are adapted to their environment. In low-input systems, this challenge is especially great, because of the challenging environment with seasonal feeding shortages, specific parasites and diseases as well as the multiple roles of the animals.

In more specialized systems, more productive traits are required besides the need for robust and healthy animals (Table 10.6).

The recommendations related to local breeding and selection in more specialized livestock keeping are divided into two parts: (i) use and choice of breeds; and (ii) breeding management.

Use and choice of breeds

The strategies for the use of choice of breeds in low-input livestock keeping are described in detail in Chapter 9, and can be maintained and improved in more specialized systems:

- Breeding selection on the basis of local criteria;
- Use of recommended local breeds.

Additional recommendations for more specialized systems are presented below.

Genetic improvement of local breeds through selection

Numerous valuable local animal breeds have been developed over time in the context of seasonal feeding shortages, specific parasites and diseases as well as the multiple roles of

Fig. 10.10. Correct spraying with insecticides – either commercial or made from medicinal plants – reduces the external parasite incidence.

Table 10.6. Comparing objectives and recommendations for improved breeding and selection in smallholder low-input and diversified livestock keeping (with more specialized animal husbandry).

Breeding	Objectives	Use of breeds	Breeding management
Low-input and diversified systems	Maintain important local breeds Make use of important traits of local breeds Effective selection Prevent in-breeding		
Recommendations for improvement		Breeding selection on basis of local criteria	Prevent inbreeding
		Use of improved local breeds	Timely castration
			Change males before mating with offspring
More specialized systems	Increased productivity Effective selection Selective use of exotics Good reproduction rates Maintain local breeds for crossbreeding		
Recommendations for improvement		Improved local breeds	Breeding only at minimum age and weight
		Crossbreeding between 25% and 75% of exotic genetics	Strict control of uterine infections
		Selection of bulls	Effective heat detection (with artificial insemination)
		Selected use of artificial insemination	

the animals. The myth of the 'unproductive' local breeds is a hard-to-crack notion, because indeed local breeds are usually less productive in conventional terms than the so-called exotic or high-yielding breeds.

These local breeds can be further selected and improved in terms of improved growth rates, improved milk production and other marketable traits. These improved local breeds can stand at the basis of more specialized animal keeping. This, however, requires a special process of breeding and selection. Most livestock projects promoting more specialized livestock keeping opt for another – and more readily available alternative introducing exotic breeds or crossbreeding exotics with animals of the local breed. This has both advantages and disadvantages.

The first step in a breeding programme is to look over the individuals of the local breed and make selection decisions on those strengths and weaknesses of males and females. This is best done with direct support of the livestock owners themselves (Fig. 10.11). Experience shows that this kind of genetic improvement through effective selection can give rise to exceptionally productive local breeds that still maintain their innate resistance and other locally valuable traits.

There are numerous examples of this practice throughout the world – though not all examples are well documented. One example is the experience of local cattle breeds from India (Gir and Kankrej from Gujarat, and Ongole from Andhra Pradesh)

Fig. 10.11. The Deccani sheep of the Deccan Plateau in India were effectively selected and improved by the communities directly involved in breeding them. Credit: ANTHRA.

that were imported into Brazil in the 1960s. Besides producing meat, these breeds were developed as excellent milk breeds after a process of selection. In fact, the world's best Gir cows today live in Brazil and give around 5500 l of milk on average per lactation. Comparing these with the neglected cousin in India, which do not yield more than 980 l, the Brazilian Gir yields roughly six times more (Sharma, 2010).

Other well documented examples are the improvement of the Aseel backyard chicken breed with support of ANTHRA in the Andhra Pradesh region of southern India (Ghotge and Ramdas, 2007) and the improvement of local Tzotzil sheep by the Indigenous Institute of Chiapas University (Perezgrovas, 2006).

This usually requires an exceptional support effort in combination with effective local organization. Sometimes the improvement of local breeds is being done in research or formal breeding institutes. This runs the risk of selecting traits that are not of major importance to the population involved.

Bringing in exotic breeds through crossbreeding

Faster changes are often desired, however. This can be done with the introduction of exotic breeds utilizing artificial insemination (AI) techniques or the import of new males, often called exotics. Many livestock development projects promote crossbreeding local breeds with exotics, with the primary objective of improving the productivity of next generations of offspring. Male animals with superior characteristics in terms of productivity are bred to local females. This can result in improved growth rates, improved milk production and other marketable traits. It is not without complications, though, because of the need for improved nutrition, some changes in management and facilities and possible decrease in innate disease resistance and ability to survive in stress conditions in the offspring.

Though this is not always implicitly recognized as such, this option implies a change from low-input into a more specialized animal keeping system. Any crossbreeding programme requires a careful consideration of the advantages and disadvantages of co-mingling the genetics of quite different breeds or family lines and the acclimatization to new environments for the offspring (Fig. 10.12). The greatest danger in using exotics lies with the ability – or inability – of the offspring to survive under conditions to which the exotic is not accustomed. Another factor to consider is that larger exotic animals may be more difficult to handle than local animals,

Fig. 10.12. When adequate management is guaranteed, local breeds can be effectively crossed up to 75% with exotic breeds, such as Jersey cattle.

or may have a size too large for regular sale on the local market.

Breeding management

Prevent inbreeding

Many livestock keepers have effective knowledge and practice related to breeding. However, it is common to find inbreeding, in both low-input and more specialized livestock keeping. Related animals breed among themselves, gradually resulting in a genetic degeneration. It does not cause mortality directly, but can lead to deformed, weak and poorly growing animals that are more likely to die.

Depending on the species, there are practices to deal with this problem:

• Selection of breeding males and castration of all other male animals;
• Timely castration of males, so they cannot mate with their mother/sisters;
• Regular exchange of reproductive males;
• Dividing animals in age groups.

Artificial insemination

The transfer of frozen semen from selected exotic males using AI techniques is routinely practised in cattle and some other species. Frozen semen can be transported over long distances in special shipping tanks and, if liquid nitrogen is available to maintain the frozen condition of the materials, the use of exotic bulls can take place almost anywhere in the world. Transferring fresh (not frozen) semen is also practised in some areas and with some species, but requires clean handling facilities and having both the male and the females in relatively close proximity. Pregnancy rates from AI are normally lower than with natural breeding.

There are many factors that go into having a successful AI programme, so the decision to go into this breeding practice must be considered in all aspects. Often, the lack of or interruption of supply of liquid nitrogen is a major limiting factor. AI technicians must be well trained and the basic equipment must be in place and in good condition. Heat detection of the animals to be bred is an art to be learned by the herd owner. Good records kept by the farmer for birthing dates and other heat dates are necessary. Females must be in a good nutritional plane and have healthy reproductive tracts. A decrease in any one of these factors will result in zero offspring or only a few pregnancies, at best.

Even with an excellent AI programme, there is the need for a male animal to breed the females that do not conceive to AI attempts. Thus, even though the costs and complexities of keeping males are decreased, they are not eliminated.

Advantages of artificial insemination:

• Genetic improvement through cross-breeding by use of selected quality semen;
• Decreased costs of feeding and housing male animals;
• Prevention of infectious reproductive diseases through natural mating;
• Documented pedigrees.

Disadvantages of artificial insemination:

• Lower fertility rate;
• Requires time and special skills of farmer for heat detection;
• Requires technical and communication infrastructure, transport and specialized input of inseminator;
• Requires technical infrastructure for semen collection, storage and transport of semen.

7: Improved Protection and Housing

Efficient protection and housing is a central element to reduce animal mortality in low-input as well as more specialized animal keeping (Table 10.7, Fig. 10.13). No major constructions are necessary – efficient constructions based on local materials have been developed by livestock-keeping families. A world can still be gained in this respect, however, as mortality because of

Table 10.7. Comparing objectives and recommendations for improved protection and housing in smallholder low-input and diversified livestock keeping (with more specialized animal husbandry).

Protection	Objectives	Predators, accidents and theft protection	Weather protection
Low-input and diversified systems	Reduced loss because of predators, theft and trampling Effective low-cost construction with local materials Prevent transmission of zoonotic parasites		
Recommendations for improvement		Protection of young animals during first weeks Protection during brooding and caring for young Night shelters Prevent cross-contamination between animal manure and human waste	Provide simple night shelters Trees for shade in fields
More specialized systems			
Recommendations for improvement		Milking shed, ventilation, manure pit management	Housing for zero-grazing

Fig. 10.13. Effective and low-cost housing for goats, which facilitates the re-collection of manure for crop production.

predators, theft and trampling is excessively high in low-input livestock keeping. This is especially the case amongst young animals.

In more specialized livestock keeping, housing is also used in zero-grazing systems and has an additional benefit of reducing the contamination of milk and other products. Moreover, effective protection in simple constructions can reduce the risk of transmitting animal diseases to humans.

The recommendations related to protection in more specialized livestock keeping are divided into two parts: (i) effective protection; and (ii) zero grazing systems.

Effective protection

The strategies for protection against predators, accidents, theft and adverse weather conditions are described in detail in Chapter 9, and can be maintained and improved in more specialized systems:

- Protection during the first weeks of life;
- Protection during brooding;
- Simple birthing pen, for example for piglets;
- Shade and wind breaks in the field;
- Special night shelters.

Additional recommendations for more specialized animal keeping are described below.

Milking sheds

Milk is often called *nature's perfect food*. However, milk is also an excellent medium for growth of foreign organisms like bacteria. In more specialized milk production units it is favourable to use sheds that are designed to reduce insect pests, and to get clean milk rapidly from the mother into a cooler or fermenting jar. Milk hygiene is critical for making good products and getting the products to market.

Ensure that the milk facility and equipment are clean and meet local milk ordinance standards for production type. Those doing the milking practice also need to maintain good personal hygiene. The shed also needs to be constructed so that the animals can move easily from the loafing/eating areas to the milking areas. Raised and solid floors help to keep moisture from collecting under foot and make it easy to remove manure.

Ventilation

Fresh air is important in all-weather types. It is unfortunately a practice in many cold regions to house livestock in a closed barn in the winter with the doors and windows closed and with little fresh air ventilation. The build-up of ammonia gases is almost overwhelming in this kind of more specialized livestock keeping.

Even high-producing dairy cattle can stand the cold relatively well and can live comfortably in temperatures well below our own level of tolerance. As long as animals can be out of standing water and out of the wind, they live very well in colder climates. Closed barns should have moving ventilation through strategically placed open windows and doors that draw a slight breeze through the facility.

Zero-grazing systems with fodder crops

Zero-grazing systems

Zero-grazing is one of the more developed practices in more specialized systems of livestock management for smallholders. It means keeping animals in a stall, and bringing fodder to them instead of sending them out to graze over large tracts of land. It is also sometimes called 'cut-and-carry'. With dairy animals, for example, it is a system that produces more milk from a small amount of land or from a rangeland grazing style of feeding.

Zero-grazing systems can be used with any of the grazing species, such as cattle, goats and sheep, and even with chickens, pigs, rabbits and other animals kept for production and marketing. Animals can be housed in an appropriately sized and simple shed with a slatted wood or hard dirt floor all made from local materials. If slat floors are used, the shed is raised about 1 m above the ground.

In other zero-grazing systems, the animals are tied to a rope, or can move around in a larger shed or open space with a simple roof building.

The major advantages of zero-grazing are:

- It saves the energy of the animal for production and reproduction that it otherwise would use for walking to and through grazing lands;

- Reduced number of pests, especially ticks and intestinal worms;
- Reduced loss of animals because of extreme weather, theft and predators;
- More land is available for growing fodder;
- Damage to crops by grazing cattle is reduced;
- Animal droppings that fall through the slats onto the ground can be collected easily for composting or direct fertilization of crops;
- Manure can be used for biofuel.

The major disadvantages of zero-grazing systems (Fig. 10.14) are:

- Construction of shed and planting fodder crops requires relatively high initial monetary and labour inputs;
- More day-to-day labour to grow, cut and carry the feed, and to fetch water;
- The maintenance of the shed and removal of manure requires extra labour;
- Lack of adequate exercise and availability of sunlight.

Requirements for zero grazing are primarily building materials, many of which come from local sources. For more extensive systems, some commercial source of lumber, cement, sand, gravel, posts and roofing material could be used. There is also the need for natural fodder or fields to grow fodder. Shed designs can be simple or complex. Basic requirements are an adequate perimeter to keep animals from breaking out, a roof for protection of all animals from the sun, a floor that can be easily cleaned and that drains away from the areas where animals lie down and stand and adequate space for feeding as well as a constant water supply.

What size? In the case of cattle, individual cubicles of about 120 cm wide × 210 cm long (4 feet × 7 feet) are best. Goats, sheep and pigs need approximately 2 m² of living space when confined to a pen.

Additional special features can be built into the zero-grazing sheds with separate sections for pregnant animals or the newborn, milking areas, larger animals kept from smaller animals and male animal segregation.

Fig. 10.14. Zero-grazing or semi-zero units that can be used for various animal species.

An adjacent paddock can be provided for animals to get out of the pen and move around.

Additional best practices for zero-grazing units:

- Provide as much water as the animals want to drink.
- Hang mineral blocks from the roof for animals to lick. Alternatively, powdered minerals can be placed in wooden boxes with an open top that are securely fastened in the feeding area.
- Cut fresh grass every day to feed to cattle. Make sure the feed troughs are never empty.
- Protect the stall from predators that are common to livestock in your area.
- If the region is very hot, extend the roof to provide a larger shaded area for cooling the air – but do not block natural air ventilation through the unit. This helps to dry manure and urine and reduce odours.
- If necessary, put straw in the pens as bedding for the animals.
- Make sure the animals have enough space for lying down, getting up and walking around. Keep the stall clean.
- Watch carefully for pests and diseases, and treat them early.
- Keep male animals separate but close to the females. Heat detection is easier

when male animals are in close proximity to the females.
- Roadside grasses tend to have more pests such as ticks, and diseases such as foot-and-mouth disease. Avoid feeding this forage to the extent possible.

8: Improved Special Care

Special care is another effective way to reduce animal mortality in all kinds of livestock-keeping systems. Special care is especially relevant around birth and for individual animals with a disease or other problem (Table 10.8). In this way, the foundation is laid for a future healthy and productive life. This only requires special knowledge and attention, without the need for expensive constructions.

The recommendations related to special care in more specialized livestock keeping are divided into two parts: (i) special care for sick animals and disease prevention; and (ii) special care around delivery.

Special care for sick animals and disease prevention

Ways to provide special care for sick animals in low-input systems is described in

Table 10.8. Comparing objectives and recommendations for improved special care in smallholder low-input and diversified livestock keeping (with more specialized animal husbandry).

Special care	Objectives	Sick animals	Around delivery
Low-input and diversified systems	Increased survival rate of sick animal Reduced mortality of newborn Reduced disease of females after birth Good bonding Reduced disease transmission of newborn		
Recommendations for improvement		Separate sick from healthy animals Shade, water, fresh feed Ethno-vet treatment Disposal of dead animals	Have animals nearby at birth Attend birth when necessary Check afterbirth Assure colostrum consumption
More specialized systems	Effective control of highly productive animals Young stock in good condition		
Recommendations for improvement		Special care of young stock Regular mastitis control Ethno-vet/commercial treatments	Support to birthing difficulties Milk fever prevention and treatment

detail in Chapter 9, and can be maintained and improved in more specialized systems:

- Isolate sick animals;
- Provide shade, fresh water, feed, and necessary treatment;
- Disposal of dead animals.

Additional recommendations for more specialized animal keeping are described below.

Regular mastitis control

Mastitis – or udder infection – is of special importance in more specialized animal keeping. Regular mastitis prevention activities can be included into milking practice.

Special care before, during and after delivery

Ways to provide special care for sick animals in low-input systems is described in detail in Chapter 9, and can be maintained and improved in more specialized systems (see below).

Extra assistance around delivery

This is especially important in more specialized systems – especially when AI is used. In this case, the young may be too large for a simple delivery. In the case of heifers (first delivery), it is better not to use AI for this reason.

Facilitate feeding with first milk (colostrum)

First feeding of milk needs to be assured within a few hours after birth. This milk with special antibodies – also known as colostrum – has special relevance to prevent disease in the first year of life. In order to have this quality it needs to be consumed during the first hours of life (Fig. 10.15).

Be prepared to treat milk fever

Milk fever is a common problem in more specialized dairy systems, which causes death within hours if not treated effectively. Because of calcium shortage in the blood, the animal cannot stand and feels very cold. A simple calcium infusion together with a follow-up treatment can save the animal. Special feeding adaptations during pregnancy can be taken to prevent this problem.

Check afterbirth

In both low-input and more specialized animal keeping, problems with the afterbirth can cause serious problems that need to be taken care of effectively.

Fig. 10.15. All newborn calves need their first milk within a few hours after birth as it contains important antibodies to prevent diseases. This is even more important in calves of the exotic breeds.

References and Further Reading

Be.Troplive (2010) Invitation Symposium 'Where medics and vets join forces', 5 November, Institute of Tropical Medicine (ITM), Antwerp.

Bhandari, D.P. (2009) *Community Animal Health Worker Manual*. Heifer International, Little Rock, Arkansas.

Birmingham, M. and Quesenberry, P. (2007) *Where there is no Animal Doctor*. Christian Veterinary Mission, Seattle, Washington.

FAO (2010) *Background Note for An International Consultation on Integrated Crop-Livestock Systems for Development – The Way Forward for Sustainable Production Intensification*. Agriculture & Consumer Protection Department in collaboration with Embrapa, IICA and IFAD. FAO, Rome.

FAO and OIE (2010) *Guide to Good Farming Practices for Animal Production and Food Safety*. FAO, Rome.

Flintan, F. and Cullis, A. (2010) *Natural Resource Management Technical Working Group, Ethiopia. Introductory Guidelines to Participatory Rangeland Management in Pastoral Areas*. Save the Children-US/FAO, Rome.

Ghotge, N. and Ramdas, S. (2007) Summary of proceedings of the National Seminar on the Sustainable Use and Conservation of the Deccani Sheep. Proceedings dated 20-22 February 2007, Hyderabad.

Hooft, K. van't (2004) *Gracias a los Animales Crianza Pecuaria Familiar en America Latina, con casos de los Valles y el Altiplano de Bolivia (Thanks to the Animals – Family Level Livestock Keeping in Latin America with Case Studies from the Bolivian Valleys and Highlands)*. Plural Publishers, La Paz.

Kumarasamy, K.K., Toleman, M.A., Walsh, T.R., Bagaria, J., Butt, F., Balakrishnan, R., *et al.* (2010) Emergence of a new antibiotic resistance mechanism in India, Pakistan and the UK: a molecular, biological, and epidemiological study. *The Lancet Infectious Diseases* 10, 597–602.

Perezgrovas, R. (2006) Direct involvement of indigenous women in sheep improvement research in Chiapas, México. Prize winning poster in German Tropentag 2006, Bonn: Prosperity and Poverty in a Globalized World – Challenges for Agricultural Research.

Sharma, D. (2010) Holy cow, Acclaimed abroad, despised at home (part 1 and 2) Blogpost on Arise India Forum. www.ariseindiaforum.org/cow_protection.php

WHO (2005) The control of neglected zoonotic diseases – a route to poverty alleviation. Report of a Joint WHO/DFID-AHP Meeting, September. WHO, Geneva.

11

Finding Pathways to Markets

Learning Objectives: Understanding
- The role of marketing in low-input and diversified systems
- The potential of niche marketing of local breeds
- The effect of marketing on female livestock keepers
- Effect of marketing on environment and culture
- Niche marketing of local breeds
- The role of marketing in more specialized systems
- The role of producer groups
- The role of local means of banking
- Cost-saving and labour-saving practices and ways of doing the work, which improves the bottom line of the enterprise

Marketing is an issue that is relatively less important to low-input and diversified livestock keepers than to the more cash-oriented, more specialized livestock keepers. However, the improvement of people's livelihoods will be very dependent on a system of markets where they can bring their animals – and the products thereof – to sell and buy. This can include both formal and informal marketing, and make use of different forms of monetary and non-monetary exchange (Fig. 11.1; Table 11.1).

There are many constraints to marketing, including distance from markets, communication, knowledge of prices, quality of the products, seasonality of the product supply etc. Where village markets occur, for example on particular days of a week or month, the distance of these markets from urban centres has an impact on the probability of traders coming to the markets. Similarly, an improved all-weather road network can make a significant difference.

Marketing in Low-input and Diversified Livestock Keeping

In most low-input and diversified systems, the livestock keepers raise their animals mainly or partly for subsistence: they or their families and neighbours consume much of the meat and milk produced, and may weave the wool into various handicrafts and garments for home use. Most also produce an unprocessed, low-value product (unsorted, unwashed fleeces;

Fig. 11.1. Local poultry being transported to the market in Guatemala. Chickens and eggs are commonly marketed livestock products which especially benefit women livestock keepers.

hides; live animals and milk) for sale. These items compete with similar, often superior products from other breeds (white Merino wool) or locations (cashmere from China, imported milk from Europe). Few livestock-keeping groups try to exploit the specific characteristics of their breeds commercially. For this and other reasons, many of the local breeds are in decline (Mathias *et al.*, 2010).

Most of the animals used in low-input and diversified systems are multi-purpose: they produce various other products and services – milk, tillage, dung and transport. In several cases, the animals are not the main source of income or livelihood for the livestock keepers; they may keep several animal species and grow crops. In this environment, the development of niche markets using the traits of local breeds is a viable option.

Poverty alleviation and economic development are usually the main motivation for development organizations to start such activities. Others aim to establish a profitable business. Additional motivations may be breed conservation, nature conservation and preserving a lifestyle.

Niche marketing of local breeds

Breed improvement and marketing programmes in the past century have concentrated on the 'big five' – cattle, sheep, goats, pigs and chickens – and breeding for production. Locally adapted breeds of these species and other, 'minor', species such as camels, donkeys and yaks have been regarded as unproductive and uneconomic, and received little attention.

Table 11.1. Comparing objectives and recommendations for improved marketing in smallholder low-input and diversified livestock keeping (with more specialized animal husbandry).

Improved marketing	Objectives	Informal markets	Formal markets
Low-input and diversified smallholder systems	To feed family and local community Improve social relationships To utilize local resources fully	Barter and other forms of exchange Sale at the doorstep	Develop niche markets of local breeds
Recommendations for improvement		Support local innovation in production, storage and marketing Improve product quality and volume Develop niche markets Improved organization	Improve product quality and volume Improve communication about market opportunities
More specialized smallholder systems	To maximize production volume Uniform quality To meet consumer desires		
Recommendations for improvement			Improve product quality and volume Develop niche markets for organic, or environmentally sound products Develop selling and buying cooperatives that provide uniform delivery through high and low seasons

A small but growing number of initiatives have started to explore the special characteristics of locally adapted livestock for economic development. A recent book *Adding Value to Livestock Diversity* (LPP, LIFE Network, IUCN–WISP & FAO, 2010; Mathias *et al.*, 2010 – see also Chapter 11, this volume) describes and analyses eight such cases – three each from Asia and Africa, and two from Latin America – where people in marginal areas produce and market specialty products from local breeds and minor species: Bactrian camels, dromedaries, goats and sheep. The raw products marketed included wool, cashmere, meat, hides and milk.

Most of the cases tried to market already existing products, either through labelling an existing product and selling it in an existing market (called 'market penetration' according to Ansoff, 1957) or through developing new markets ('market development'). None of the cases started with the development of new products for existing markets ('product development'), although some did so at later stages. Three diversified their product spectrum through developing new products for new markets ('diversification').

The 'four Ps' of fundamental issues that marketing initiatives need to address are Product, Price, Place and Promotion. In all cases, these issues were addressed and the livestock keepers benefited through higher, more stable prices, increased demand for the products of their local breeds, a more reliable market or some combination of these items. These outcomes were reached through enhancing the quality or the amount of the raw material or creating a market for it. The livestock keepers also benefit in another, more intangible way. By becoming part of a value chain that increases their incomes without damaging the environment, they can gather government support. This is important in areas where governments tend to view livestock keepers, especially itinerant ones, as a problem or threat, and try to get them to change their lifestyles, settle in permanent locations and start growing crops.

Type of interventions

The projects for niche marketing of local breeds usually focus on four different types of interventions (Mathias *et al.*, 2010).

Animal production

Many projects attempt to increase or improve production of the animals that produce the raw materials: by establishing breeding herds, increasing the number of animals with the desired traits, and improving animal management and health. However, in working with local breeds, the production of the raw material is usually not a major focus of the project.

In low-input and diversified livestock keeping the projects usually do not focus on modifying animal production to achieve specific production standards (such as organic production) or other production-related goals (e.g. environmental and breed conservation, or enhanced animal welfare standards), but such goals are often indirectly included.

Processing

Improving the processing of the raw materials is often a major focus in developing projects related to niche marketing of local breeds. This can include establishing factories, designing new products, introducing new techniques, and improving sorting and grading.

Organizing

Organizing groups of producers and processors is key in most marketing development projects. This might mean organizing community members in production cooperatives, employing them as staff, establishing formal companies or subcontracting work out to processors. Organizing efforts are not always successful, however.

Building a value chain

All of the cases included efforts to identify markets and build a value chain, linking producers with processors and markets.

Target markets

Most marketing development projects aim to target a specific market. In the case of niche marketing, this is often aimed at producing specialty products targeting environment-conscious consumers, tourists, fashion houses, hobbyists and barbecue party hosts in urban centres. Three of the seven projects export their products (sheep wool in India, cashmere in Kyrgyzstan, camel wool in Mongolia). None focuses primarily on local rural consumers.

Marketing and women livestock keepers

Marketing of products from local breeds and minor species offers benefits for women, especially if the products are fibre or milk based. Women are directly involved in all the cases in various activities: production, processing and marketing.

Women and men often play complementary roles in livestock raising: men typically manage the larger animals (cattle, camels), shear the wool and sell high-priced assets such as livestock. Women typically are responsible for smaller animals (sheep, goats) and calves, and handle activities such as spinning and weaving, and sell low-priced products such as milk and wool. This division of labour is most clearly shown in the Somalia case.

This distinction opens the possibility for value chains to empower women and benefit them economically. Women in our cases earned income, learned skills, and gained power and respect in their societies. They also invested significant amounts of time and effort in work that can be tedious (spinning), physically demanding (hauling heavy milk cans) or hazardous (travelling long distances). They are forced to balance this work against other demands on their time, such as childcare, household work and managing livestock. Their other commitments may limit their incomes from the marketing activities. More women might benefit if equipment could be introduced to reduce drudgery – though the introducing machinery sometimes means a shift in tasks and benefits to men (Mathias *et al.*, 2010).

When we think of ways to increase women's participation in livestock markets, we usually think of increasing their ability to sell animals or animal products. While this can be an important way to improve the welfare of women and their families, the issues raised in previous sections make it clear that, unless women are able to make decisions about which products and animals are sold and what is done with the proceeds of the sale, increasing market participation alone may not benefit women (Kristjanson *et al.*, 2010).

The actors in livestock value chains include not only livestock producers but also input suppliers, traders, processors, wholesalers and retailers. Helping women gain access to labour, product and service markets all along the value chain, and improving their working conditions, are additional ways in which women can benefit from participation in livestock markets. While women may play many of these roles along the value chain in many regions, the literature mainly cites their roles as suppliers of livestock products, particularly milk products, and as processors of animal source foods, often street foods.

There appears to be more awareness of the importance of gender in market-related livestock projects than in projects focused on raising livestock productivity. Whether this awareness translates into effective livestock marketing strategies for women is unclear. A Heifer International report on activities in East Africa found that women provided more labour in dairy enterprises than men, but the level of women's control of the dairy income did not usually match their contribution. This was in spite of Heifer's finding of a strong correlation between women's control of dairy income and the productivity and success of dairy projects (Heifer International, 2008).

Women's groups initiated by development projects are widely used to support women pursuing urban agriculture; these groups provide women with micro-credit schemes and other forms of support for their dairying, poultry production, livestock marketing, and food transformation and sale (Niamir-Fuller, 1994; de Haan, 2001).

Joining such groups may be the only way for many poor women to obtain sufficient resources to start up and profitably operate a livestock-related enterprise (Fig. 11.2). Membership in such groups enables women more effectively to lobby government departments and other decision-making agencies affecting their livelihoods. Although the performance of such women's groups has been reported as variable, group membership gives many developing-country women the freedom to participate in livestock development activities, enabling them to protect their interests, to overcome legal hurdles facing them, and to access the training and equipment they need to increase their production and sale of safe livestock foods.

The scarce literature that exists on women and livestock markets indicates that developing country women participate in livestock value chains mainly as suppliers of dairy products and as producers and sellers of processed animal-source foods in informal markets. Although increasing the participation of women in livestock markets and value chains clearly has the potential to improve welfare, the increasing commercialization of livestock markets presents women with risks as well as rewards. The literature cites many cases where women's control over livestock enterprises and incomes is diminished rather than maintained or enhanced with increasing commercialization.

Fig. 11.2. In low-input livestock keeping, women can be involved in various elements of the livestock value chain. Supported by VSF, women produce and sell animal health products based on medicinal plants in northern Guatemala.

Women stand to benefit substantially from improvements in food safety, especially in informal markets, but are often inadvertently hurt by the unintended consequences of inappropriate policies and regulations. The conditions leading to these different outcomes need to be much better understood. While market-oriented livestock projects, perhaps more than productivity-focused projects, are increasingly recognizing the need to pay attention to gender, the challenge remains to identify strategies that help women enter into and benefit more from livestock markets (Kristjanson *et al.*, 2010).

Effects on environment

Linking livestock keepers to a value chain may have adverse effects on the environment. This may occur if the owners begin to keep more animals than the environment can sustain.

Adverse effects may also occur if the mobility of pastoralists is constrained.

In Somalia, a more flexible system has emerged, where the milk collectors – themselves community members – follow the herds during the migrations. This enables and encourages mobility. Even here, though, some herders have begun to keep their lactating camels near their huts where they can milk them easily, while sending non-lactating animals further afield.

Broader trends may mask or accentuate the environmental effects of a marketing project. In most countries, rapid urbanization, population growth, changing lifestyles, the conversion of land to other uses, the decline of mobile pastoralism and climate change are much larger influences on the environment than the creation of a value chain for a particular product. In Mauritania, for example, it is unclear whether the tendency for pastoralists to settle in one location is because of dairy's milk purchases or part of a broader trend towards settlement and urbanization. While the causes for such changes lie outside its control, a marketing initiative can reinforce them, accept them as a fact of life or try to counteract them (Mathias *et al.*, 2010).

Livestock marketing and culture

Marketing efforts related to local breeds can both undermine and reinforce local culture. For example, empowering women, trivializing traditional products in order to please tourists, opening contacts with a consumer society, or encouraging mobile herders to settle in one place potentially can weaken the local culture. The outcome of such changes lastly depends on how the society handles them. If empowering women leads to increased divorce rates as reported in the Mauritania case, the women will be better off only if the traditional or national laws do not outcast or disadvantage divorced women.

In other instances, the marketing efforts can reinforce the local culture, for example, by increasing the awareness and pride of local people and outsiders in their cultural values (including the local breeds), empowering local people to press for their interests, encouraging them to rediscover lost skills or reviving traditional handicrafts (Mathias *et al.*, 2010).

In the marketing related to many low-input and diversified systems, the monetary market systems are often combined with culturally defined non-monetary forms of exchange. This is also known as reciprocity economy. One example is the Seventh Friday Fair in Sipe Sipe, Bolivia, where once a year farmers from highlands and valleys exchange their products and strengthen their social relations (Box 11.1).

Marketing and sustainability

How sustainable are the marketing initiatives related to local breeds? Four of the eight cases (Mathias *et al.*, 2010) appear to be sustainable. In India, the wool enterprise has a profitable business model, a growing pool of suppliers and long-term relationships with its buyers. The poncho makers in Argentina appear to be serving a niche, though it is unclear how large its potential market is and whether it can grow significantly. In Mauritania, the Tiviski dairy is the market leader; it has a long history of creating innovative products and successfully competes with lower-priced rivals. The discovery that camel milk has therapeutic qualities is opening up a promising niche market of diabetic or health-conscious consumers. In Somalia, the women traders supply a rapidly growing urban market with a vital product (Mathias *et al.*, 2010).

Box 11.1. Reciprocity economy in Seventh Friday Fair.

In the Andean marketplaces and smaller 'ferias', the Andean reciprocity economy and the monetary marketing realities are combined. Here you can find both monetary exchange as well as various forms of non-monetary exchange.

For example, on the seventh Friday after Easter, a fair takes place in the small town of Sipe Sipe near Cochabamba, Bolivia, during which farmers from different agricultural ecosystems exchange their recently harvested produce. Farmers from the highlands bring various varieties of potatoes and of other local tubers, like oca and papalisa. Farmers from the valleys bring local varieties of maize. The abundant biodiversity is clearly reflected at this colourful fair, but there is much more to it. The fair is a clear example of the strength of the 'reciprocity economy'.

In spite of centuries of Spanish colonization, and the total hegemony of the monetary market system today, this system of reciprocity economy has survived amongst indigenous groups in many parts of the Bolivian Andes. These socio-economic relations of production have allowed them to maintain their principles of community and solidarity, within a system of exploitation (Compas and Agruco, 2010).

That does not mean that these initiatives are secure. Foreigners' tastes for Indian handicrafts may change; a recession in Argentina may mean fewer tourists with less money to spend on ponchos; subsidized imports from the EU may ruin Mauritania's dairy sales; civil war may disrupt the Somali milk traders, or a reinvigorated government may introduce taxes or hygiene and veterinary controls, but these are risks similar to those faced by many businesses, and not just in the livestock sector or in the developing world.

The future of the other four enterprises is more doubtful. The Kyrgyzstan goats initiative shows promise: it is based on an existing resource and is not capital-intensive. However, it depends on transferring knowledge and skills, establishing a reliable value chain, and building strong local institutions. It is also sensitive to the world price for cashmere and the activities of Chinese traders in the country. Government support is needed to ensure that this chain can become better established.

In Mongolia, the camel wool initiative must make the difficult jump from a project-sponsored activity to a self-sustaining business venture. It is necessary to nurture local institutions that can coordinate the wool production and marketing. Without this, the enthusiasm of the donors and volunteers will eventually wane, and local people will be unable to take on their roles.

In South Africa, the goats enterprise must overcome governance problems and ensure a reliable supply of live animals so it can expand its operations. This will probably mean putting more emphasis on its commercial operations rather than its social responsibilities. This is a large project, so it is in the interests of the government, its main sponsor, to ensure that its money has been invested wisely.

The Criollo goats initiative in Argentina is too new to judge whether it will be a success. As the first application under the law that governs the country's Protected Designations of Origin, it is charting new territory. Much will depend on whether consumers can be persuaded to pay extra for a specialty product, whether the board that manages the designation of origin functions as hoped, and whether livestock keepers can benefit financially from the labelling. An additional risk is competition: if Argentina's many other meat producers see it as a successful marketing effort, they are likely to imitate it, driving down prices and eliminating any financial benefits for the Criollo goat keepers (Mathias *et al.*, 2010).

Marketing and wealth disparity

Some examples also show that increased marketing may also imply an increase in wealth disparity. Markets are often outside the span of control of the livestock keepers themselves, especially when the drivers of market demand are from outside the region. This is the case in the pastoralist marketing in the Somali region of Ethiopia (Box 11.2), which shows that the effect of improved marketing opportunities may increase wealth disparity between poorer and richer households (Akililu and Catley, 2010).

Recommendations for supporting marketing in low-input and diversified livestock keeping

What elements are needed for a marketing initiative based on local breeds and minor species to be successful and sustainable? Here are some suggestions (Mathias *et al.*, 2010):

- *Use existing resources.* The initiative should be based on existing resources: the livestock breed, natural resources, traditional knowledge and human resources, and use the environment in a sustainable way.
- *Identify a suitable entry point.* To conserve a breed or benefit livestock keepers, it may be better to focus on some aspect of the chain other than working directly with livestock keepers. For example, developing an urban-based processing industry to increase demand for the raw materials may be the best way to benefit livestock keepers (or conserve the breed).

Box 11.2. Increased pastoralist marketing opportunities in the Somali region of Ethiopia.

The Somali Region has a long history of livestock exports, especially live animals channelled into the cross-border trade to Somaliland and Puntland, and then onwards to the Middle East. Dating back to the 1920s or before, this trade is both robust and growing as demand for meat increases with urbanization, population growth and affluence in the Gulf. More recently, and with government support to formal meat exports, Borana pastoralist areas have been supplying increasing numbers of livestock to export abattoirs, but who benefits from these trends, specifically, in pastoralist areas?

The answer lies partly in an understanding of wealth stratification among pastoralists, and the differing strategies used by poorer and richer households to build and maintain financial capital, i.e. livestock. In general, poorer households must prioritize the building of herds if they are to acquire sufficient numbers of animals to withstand shocks and droughts. This strategy, despite its inherent economic logic, also limits the extent to which they can or should sell animals. In contrast, richer herders are the main suppliers for livestock export markets. These herders already have sufficient animals to better survive drought, and have excess animals to sell. Furthermore, as wealthier households benefit from sales they also have greater capacity to control key land and water resources, which, directly or indirectly, have negative impacts on poorer herders. This is most evident when hitherto communal resources are 'privatized'.

The sum outcome is an increasing asset gap and a gradual redistribution of livestock from poor to rich. This trend explains why these pastoral areas can export increasing numbers of livestock, but are also characterized by increasing levels of destitution. The report estimates annual increases in the number of wealthy pastoral households of around 2.5% (in line with average population growth), but increases in poor households of 4.1% (Akililu and Catley, 2010).

- *Start small.* The initiative should invest first in human capital and at a small scale, rather than in costly infrastructure. If the activity works, it should then seek more capital investment.
- *Do the research.* It should be based on a thorough understanding of the production system, the product and the market. That means studying the breed and its characteristics, the livestock keepers and their production system, the range of potential products, and the potential customers for the products.
- *Identify special characteristics of the breed.* The initiative should seek ways to market products that reflect these characteristics: by creating new products, refining existing traditional products, or finding new markets for existing products.
- *Find a viable business model.* The initiative should generate income for all actors in the value chain.
- *Focus on quality.* It should emphasize the need to maintain quality. A specialty product can command higher prices only if it is superior to alternative products.

- *Build capacity.* It should stimulate the creation of strong local institutions and train people in technical and management skills.
- *Don't depend too much on outsiders.* The initiative may require significant support from outsiders over the medium term, but should not depend on expertise or funding from outsiders over the long term.
- *Ensure long-term demand.* The product chosen should be one where demand is likely to grow over the long term.
- *Don't put all your eggs in one basket.* The initiative should be based on a range of products and markets: that way, it is not a disaster if one product fails to sell or one customer refuses to buy.

Marketing in More Specialized Livestock Keeping

Many interventions in more specialized smallholder livestock keeping are aimed at increasing their production as well as improving the quality of products to meet the market standards. In many developing projects, farmers receive support in the form of training on the use of new technologies in production. The farmers also receive support in accessing inputs such as breeding services, feeding and animal health services. These services are provided by service providers who are paid for their services by farmers. This support ensures that the services farmers receive will continue to be provided after the intervention ends.

Access to distant markets

Access to more distant markets is another strategy for more specialized systems. Local roads and communication resources to access market information is imperative. Group formation becomes more important also. Crop and livestock produce from neighbours can be combined for volume delivery, which most often demands a higher price. Consumers who live and work away from agriculture often develop specific desires for types and quality of agriculture products. This presents an optimum opportunity to construct specific market commodities for these consumer demands.

One outside influence that can derail more specialized systems is the import of similar and lower cost commodities from outside the region. Often, this is tied up in trade agreements that are outside the control of local producers and takes combined efforts to develop protective legal structures at the national government level.

Farmers' groups

Another area of support is on institutional strengthening, where the farmers establish an institution, which will help them in organizing and operating the markets. The institutions will link up farmers to markets. They will pool farmers' produce and sell to upper markets such as dairy processors or high-end consumers. The farmer institutions

will enter into contract in marketing of farmers' produce and pay farmers when the produce has been sold (Box 11.3).

The role of Heifer in this market development comes in two key areas: development of the institution that will manage the market and linking this institution to the market. By supporting farmers in this way, Heifer is facilitating the smallholder livestock farmers to have access to the markets.

Use of standards and new products

Heifer International uses the standards set by the markets either globally or locally while supporting farmers to meet those standards. The milk quality standards are fairly uniform internationally. There could be some allowances for lower standards in developing countries but milk sold to the public must meet basic wholesome standards including low bacteria count and minimum levels of butterfat content and other solids not fat. Each country has its own local standards but milk crossing the border must meet internationally recognized quality standards.

If there are new products being introduced to the country or any local population, the quality standards remain the same as those accepted internationally (see also Kenya Dairy Board, 2005).

Farmer field schools

The farmer field school is the basic farmer institution where farmers come to learn new farming technologies, adopt them while they are learning and also obtain inputs to their farming enterprises. As the production increases, the farmers will have need for market outlets where they can sell their surplus produce.

The farmer field school can move to a higher level of legal entity before it engages itself in market organization. The field school can move to farmer association, cooperative society or limited liability

---

Box 11.3. Example of dairy development and marketing in Kenya.

Mr Laban Kipkernoi Talam is a youth by Kenyan standards. In 2002, Laban 'borrowed' a dairy cow from his relative to help him take care of the nutrition of his family and to provide some income. To meet his family's ever-growing needs, he decided to venture into a brick-making business. He also expanded his horticulture farming business to include capsicum and other local vegetables. He used to rent 0.2 acres of land for the gardens and was paying Ksh 1000 (~US$13.30) per month. In 2005, after seeing the benefits of being in a successful group in the area, he started a group of his own with 16 members called the Silanga Youth Group. The group was mainly involved in merry-go-round activities, such as horticulture and tree planting.

In 2005, the group applied for funds from the Government of Kenya Youth Enterprise Development Fund (YEDF). The group was, however, not successful because they had no bank account. In 2007, they tried again for the same funds. This time, they were successful, as the group had established a bank account. They were given Ksh 46,600 (~US$621.30). As a group, they used the money to expand their horticulture farming business from 1 to 3 acres, with each member having a portion of between 0.2 and 0.3 acres. They cleared the loan from YEDF in 2008 and used the profits that they had accumulated in their account to buy a dairy heifer for the group.

Then, a new programme was introduced into the area. The East Africa Dairy Development (EADD) programme began working with dairy farmers, such as Laban, in Kabiyet in August 2008. EADD started mobilizing farmers through training meetings. Laban and his group received training through demonstrations on animal husbandry, fodder management, soil conservation and record keeping. The group also started buying shares from the newly registered Kabiyet Dairies Ltd. The demonstrations that were held at Laban's farm included planting of Elba Rhodes, Calliandra, Kakamega, sweet potatoes, lucerne and *Desmodium*. As the trainings continued, Laban was selected by the EADD team for a Training of Trainers course. He started using his acquired skills to train his group and other Dairy Management Groups (DMGs) that were developed by EADD. Laban now trains, on average, 60 farmers per month from in and out of the county on animal husbandry.

From the EADD and Government of Kenya (GOK) extension staff training, Laban has managed drastically to change his dairy cow management practices, particularly in on-farm feed formulation. This has greatly boosted his milk production from an average of 5 to 20 l per day. It has also drastically reduced his costs of buying expensive feed supplements from the agrovet shops (see also Chapter 12, Case Study 4).

---

company, where it will have strength to enter into contract with others. Several farmer field schools in a region can form a federation, which is again of higher legal status than the basic farmer field school organization.

Additional planning

More specialized agricultural practices take greater long-term planning, more sophisticated inputs and succession planning for labour and management. In turn, it provides greater market variety, volume and geographic distribution. As described in Box 11.4, additional planning may include business plans, in which assessments are made of potential and assured market demand, improvements in natural resource inputs that yield greater returns, optimum timing of application of inputs and harvest, all with adequate labour to gather crops and livestock production at the optimum time for product quality.

Once crops and livestock products are harvested, it is imperative to have adequate storage volume and conditions, to hold product through the market season. Market planning then includes the timing of the next production season to minimize the lack of desired products before the next harvest. Some harvested products are improved during storage with specialized techniques, such as silage making and mixing of nutrition ingredients.

Often, there is the need for capital to be used for longer-term infrastructure development and other business needs. Appropriate loan products that are available to farmers and processors are needed. These are often provided through micro-finance institutions and banks.

Box 11.4. Ramat Livestock Enterprises in Kenya.

Maasai Pastoralists in Narok District, Rift Valley, Kenya are on the path towards sustainable livelihood through new innovation at the thrust of modern livestock production and marketing. With support from Heifer International Kenya, two Maasai communities comprising over 12,000 families are moving from subsistence livestock keeping to sustainable commercial livestock production and marketing. Heifer Kenya has supported the Maasai community overcome perennial challenges of loss of livestock and their livelihood from prolonged droughts, water shortages, environmental degradation, preventable livestock diseases and lack of livestock market access. To address sustainably the lack of market access for their livestock, two Maasai communities organized themselves into a business company called Ramat Livestock Enterprises in May 2007. The company (similar to cooperative movement) is now in the fore-front of spearheading modern business approach to livestock production practices and marketing together with the existing community organizations (also shareholders of Ramat Ltd) namely Loita Development Foundation and Keekonyokie Suswa Trust (KST) who are spearheading community mobilization and new livestock technological practices in their respective communities

Despite the many livestock that the Maasai community own, Narok District has had the highest poverty incidences in Kenya. The Kenya government poverty mapping report for 2003 shows that 60% of Loita and 54% of Keekonyokie people live below the poverty line respectively (or live on less than US$1 a day).

In order to advance its course, Ramat Ltd has leased 1000 acres to establish a livestock holding facil-ity from the KST community in Suswa, 80 km South of Nairobi City. The location was chosen because of the land availability and its close proximity to Nairobi City, with all potential opportunities for niche livestock markets nearby. This is despite the remoteness of the place and its dryness throughout the year with annual rainfall amounts ranging from 100 to 300 mm. The land is heavily denuded but Ramat is prepared to showcase the community how to manage denuded land and prevent soil erosion. Ramat has fenced the land all round for ease of proper management and planning. With support from Heifer Kenya, Ramat has developed several infrastructural facilities enough to justify it as a feedlot facility. These facilities include: fencing, two cattle feedlots, three water reservoirs that can hold up to 1.5 million litres of water, sheep and goat holding yards, livestock handling facilities like weigh bridges, spray race and crushes (Fig. 11.3). A total of 350 acres of pasture field has been established where various grass species

Fig. 11.3. Steers in the livestock holding ground before sale in Suswa, 80 km South of Nairobi, Kenya.

Continued

Box 11.4. Continued.

---

have been planted purposely to be harvested and stored as hay and silage to be used in the feedlot and even during dry periods. The feedlot facility can hold up to 800 cattle and 1500 sheep and goats for fattening at the same time. A total of 300 steers and 600 sheep have been fattened and sold through the facility to date since June 2009. Currently biogas facilities are being constructed to make use of livestock manure from the feedlot, produce biogas for electricity generation within the holding ground and reduce the level of methane production, which is harmful to the ozone layer.

Besides quantity, volume and quality, consumers want to know that their produce, both crop and livestock is produced in humane and safe ways. Organic and natural agriculture provides food commodities that certainly meet these demands. Other consumers want to support genetic conservation with the purchase of rare and heritage breeds of livestock and varieties of crops. These special markets are available and can demand greater return than conventional produce.

References and Further Reading

Akililu, Y. and Catley A. (2010) *Mind the Gap – Commercialization, Livelihoods and Wealth Disparity in Pastoralist areas of Ethiopia*. Tufts University, Medford, Massachusetts.

Ansoff, I. (1957) Strategies for diversification. *Harvard Business Review* 35, 113–124.

Compas and Agruco (2010) Local markets and reciprocity economy in Bolivia. Compas Policy Brief no. 4.

Haan N. de (2001) Of goats and groups: a study on social capital in development projects. *Agriculture and Human Values* 18, 27–39.

Heifer International (2008) 2007–2008 Project profiles: where we work and what we do. Heifer International, Little Rock, Arkansas. Available from http://www.heiferproject.org.

Kenya Dairy Board (2005) Improve the Quality of Your Milk and Please Your Customers. Training guide for small-scale informal milk traders in Kenya.

Kristjanson, P., Waters-Bayer, A., Johnson, N., Tipilda, A., Njuki, J., Baltenweck, I., Grace, D. and MacMillan, S. (2010) Livestock and women's livelihoods: a review of the recent evidence. Discussion Paper No. 20. ILRI, Nairobi.

LPP, LIFE Network, IUCN–WISP & FAO (2010) *Adding Value to Livestock Diversity – Marketing to Promote Local Breeds and Improve Livelihoods*. Animal Production and Health Paper 168. FAO, Rome.

Mathias, E., Mundy, P. and Köhler-Rollefson, I. (2010) Marketing products from local livestock breeds: an analysis of eight cases. *Animal Genetic Resources* 47, 59–72.

Niamir-Fuller, M. (1994) Women livestock managers in the Third World: focus on technical issues related to gender roles in livestock production. Staff Working Paper 18, IFAD, Rome.

12

Challenges and Best Practices in Livestock Development Support

Case Study 1: A Story of Rubbish and Pigs in Cairo – The Effects of Swine Flu Control in Egypt. *Ellen Geerlings*

Case Study 2: The Shepherds and their Black Sheep of the Deccan, India. *Nitya Ghotge and Sagari Ramdas, ANTHRA, India*

Case Study 3: Bio-cultural Community Protocols to Support the Samburu Red Maasai Sheep and their Shepherds. *Jacob Wanyama, LIFE Network – Africa Region*

Case Study 4: The Heifer Methodology of Supporting Sustainable Livelihood in Kenya – The Story of Mr Laban Kipkernoi Talam. *Reuben Koech, Heifer International Kenya*

Case Study 5: Ethno-veterinary Medicine in a Master's Degree Programme on Tropical Agricultural Production. *Raúl Perezgrovas and Guadalupe Rodríguez, Instituto de Estudios Indígenas-UNACH, México*

Case Study 6: The Dairy Development Trap: How Developing Countries can Learn from the Experiences of Dutch Dairy Farming. *Katrien van't Hooft, Dutch Farm Visits, The Netherlands*

Case Study 1: A Story of Rubbish and Pigs in Cairo – The effects of Swine Flu Control in Egypt

By Ellen Geerlings, researcher on the socio-economic impacts of animal disease control on vulnerable households (ellen.geerlings@ hotmail.com) (for more information please see Geerlings, 2007)

In Mokattam, an area on the outskirts of Cairo, the Zaballeen reside – a community of 60,000 people, the vast majority (95%) Coptic Christians and the remaining part (5%) are Muslim. They live as a minority group within the primarily Islamic society of Egypt. The word 'Zaballeen' means 'rubbish people'. They derive their main income from rubbish collection, rubbish sorting and rubbish recycling. The Zaballeen inhabit several areas on the outskirts of Cairo, among these are Tora, Katameya and Mokattam. Mokattam is the largest area and it is often referred to as the largest 'rubbish city' in the world.

I was curious to see the area after seeing the documentary 'Garbage Dreams' by Mai Iskander: 'Filmed over four years, GARBAGE DREAMS follows three teenage boys born into the Zaballeen's trash trade: 17-year-old

---

Box 12.1. A message about swine flu culling in Egypt (Reuters, 5 May 2009) (www.Reuters.com).

Egypt, hard hit by the highly pathogenic Bird Flu Virus, is considering culling hundreds of thousands of pigs as a precautionary measure as Swine Flu nears the borders of the most populous Arab Country. Experts fear any Flu Pandemic could spread quickly in Egypt and have a devastating impact in a country where most of the roughly 80 million people live in the densely packed Nile Valley, many concentrated in crowded slums in and around Cairo.

Egypt considering culling pigs due to Swine Flu fears, between 300,000 to 400,000 pigs could be affected. Egypt will compensate affected farmers. The move is not expected to block the H1N1 virus from striking, as the illness is spread by people and not present in Egyptian swine. But acting against pigs, largely viewed as unclean in conservative Muslim Egypt, could help quell a panic.

A spokesman for the Egyptian cabinet, Magdy Rady, reported Wednesday that 'the question now is should we kill them or relocate them, and the prevailing idea now is to kill the existing pigs and of course compensate their owners'. Rady put the number of pigs that could be culled at between 300,000 and 400,000, and said a decision was expected in days. 'If you see the conditions of the Swine farms in Egypt, they are not healthy at all. They are hazards in themselves, even without the Swine Flu. That's why people are really getting afraid.'

Egypt, harder hit by the H5N1 Bird Flu Virus than any other country outside of Asia, is deeply worried about the impact of another Flu Virus after Bird Flu inflicted extensive damage to its poultry industry and economy in 2005.

Adham, 16-year-old Osama, and 18-year-old Nabil. Laila, a community activist who also teaches the boys at their neighborhood Recycling School, guides the boys as they transition into adulthood at a time when the Zaballeen community is at a crossroads'.

I was eager to know how the Zaballeen have coped with the loss of their pigs in June 2009, when the government decided to cull them in an attempt to control swine flu (Box 12.1). The first thing I noticed when I arrived in Mokattam was that the streets are buzzing with energy: pick-up trucks and donkey carts driving off and on loaded with unsorted and sorted rubbish; people sorting through heaps of plastic bottles; men cutting cans to separate the more valuable lid from the rest of the can; women removing copper from discarded electrical appliances; plastic being shredded and washed, the list goes on and on. Beyond the dirt and chaos that is apparent at first sight one also finds a highly organized and efficient community.

The Zaballeen have developed an amazingly efficient system; more than 80% of Cairo's rubbish is recycled – according to the people to whom I spoke. On average, the Zaballeen collect about 4000 t of waste per day, using (Fig. 12.1) the high-rise pig feeding floors and other sorting floors. The efficiency

of the Zaballeen is rooted in the fact that every single piece of rubbish is handled manually and even the smallest pieces are sorted and recycled. Rubbish contains both non-organic and organic waste, the latter making up about 55% of the total rubbish collected by the Zaballeen. In order to make use of the organic waste, about 80% of Zaballeen used to keep pigs.

Then the outbreak of swine flu in 2009 moved the Egyptian government preemptively to cull all pigs of the Zaballeen. The culling was not without resistance; dreading the loss of their pigs and with that food and income, Zaballeen community members clashed with the police. Their protests were in vain and they were forced to give up their pigs. The pigs had to be brought to a specific area, which the government had designated as the culling and disposal site. Reportedly, some pigs were beaten to death or even buried alive. All houses were searched to make sure no pigs were left alive; some Zaballeen had attempted to hide their pigs to avoid culling. Sows and piglets were disposed of by burying them and no compensation was received. The Zaballeen were partially compensated for the loss of their male pigs; 40EGP was received for each one of them. These pigs were not buried but loaded on

Fig. 12.1. The non-Muslim Zaballeen in Egypt are the 'rubbish people' that derive their main income from rubbish collection, rubbish sorting and rubbish recycling. High-rise buildings in Cairo provide separate areas for keeping animals and sorting rubbish. Credit: Ellen Geerlings.

government trucks and sold according to one informant.

The total number of pigs before the culling was approximately 300,000, mostly in the hands of the Coptic Christian Zaballeen minority groups. The number of pigs in each household varied between 10 and 100, depending on the amount of rubbish they would collect (i.e. organic waste to feed the pigs) and the space available to keep the animals. The pigs provided income and a year-round meat supply to the families. Normally the biggest share of income was derived from rubbish sorting and selling, but sometimes when prices of goods would go down the pigs provided a good buffer. Depending on the weight, a medium-sized pig would usually fetch about EGP100–150 (EGP6/kg live weight).

Apart from the culling of their pigs, the Zaballeen also had to deal with new government regulations that forbid them to collect the rubbish in Cairo city. In an attempt to upscale the rubbish collection in Cairo, the government contracted big multinational companies to take care of rubbish collection. Before the multinationals came, households all over Cairo would pay the Zaballeen a small fee to collect the rubbish from their doorstep. With the coming of the multinational companies, the government imposed a monthly fee of EGP3 – added to the electricity bill – to cover part of the costs. These multinational companies only recycled around 20% of all the rubbish, only collecting the rubbish in the rubbish bins and sweeping the streets. Most multinational companies have failed and have meanwhile left the country, leaving only two companies in charge of the near impossible task of collecting all the rubbish.

Rubbish collection without a licence is now forbidden and the only way for the Zaballeen to collect rubbish is to be incorporated within the multinational companies – or risk fines when collecting the rubbish illegally. Some Zaballeen have now been integrated into the multinational waste management system. They are allowed to collect and keep the rubbish in exchange for their rubbish collection services, but do not receive any salary for their work. Meanwhile, many Cairo households have ceased to pay the Zaballeen a collection fee because they are already paying the monthly fee of EGP3. Many Zaballeen have thus been cut off from their rubbish supply and the few that have managed to work with the multinational companies find it far less lucrative now than before.

After the pigs were culled and the multinational companies came, many families had to stop collecting rubbish. Some successfully invested their money into a shredding machine and now buy plastic waste, which they then shred and wash. This has been a lucrative strategy and they have managed to make more money now than they used to before. Other families have followed this example and buy raw material by the kilo, shred it, wash it and then make items such as clothes hangers, which they sell per piece. Raw material is bought from people working on rubbish dumpsites and from rubbish collectors and scavengers.

Many other families have been less fortunate and are having a hard time, particularly as a consequence of the loss of their pigs. They do not have the means to invest in a shredding machine and have a hard time even getting enough food. In one case, the pigs provided savings for the marriage of two sons (26 and 28); now with the loss of the pigs they can no longer save – while the sons are now reaching an age where it will be very difficult to get married. This is a big burden and cause of stress in the family.

Most Zaballeen want to have pigs again but there are only very few local pigs left in this part of Egypt. Most pig farms depend on the government to import exotic pigs. Some Zaballeen families have started to keep goats and to a lesser extent sheep. These species are not as efficient as pigs in recycling the organic waste – but they make use of at least part of the organic waste, and people can earn an income from selling live animals as well as having them for home consumption.

Reference

Geerlings, E. (2007) Highly pathogenic avian influenza: A rapid assessment of the socio-economic impact on vulnerable households in Egypt. FAO/WFP Joint Project Report, Rome.

Case Study 2: The Shepherds and their Black Sheep of the Deccan, India

By Nitya Ghotge and Sagari Ramdas, ANTHRA, India (www.anthra.org)

The Deccan plateau of south central India is one of the driest parts of the country. Crop rearing is precarious and rural livelihoods are dependent on livestock to augment and supplement them. For hundreds of years, shepherding groups like the Kurumas, Kurubas, Gollas and Dhangars have traditionally reared sheep and goat.

Where possible, some of them practise agriculture during the four monsoon months from June to September. As the rains recede, these groups migrate with their animals to where fodder is available. Over time, these groups have developed strong relations with settled farming communities who trade agricultural produce in return for dung, milk, meat and wool from the shepherds. Allied livelihoods have also developed around these communities such as the spinning of wool and weaving. These different castes, groups and subgroups and their specialized livelihoods woven together form the fabric of society, which sustains thousands of households in the otherwise harsh terrain of the Deccan. Today, India ranks second in goat meat production and seventh in sheep meat production in the world, and all of this meat is produced through low-input pastoral systems such as that found in the Deccan plateau.

The Deccani sheep breed

The breeds of this region have evolved over the years to suit the ecological and social landscape of the region. The black Deccani sheep is the most popular sheep breed of this region and is reared by different pastoral and agro-pastoral communities: the Kurmas and Gollas in Andhra Pradesh, Dhangars in Maharashtra and the Kurubas in Karnataka. The Deccani is a medium-sized, short-tailed coarse wool sheep breed and black is the predominant colour though shades of tan, brown and even white are observed in some flocks. This hardy breed shown in Fig. 12.2 is ideally suited to the extreme temperatures of the Deccan and is capable of long-distance migration, a necessary coping strategy during years of drought. The breed is mainly reared for meat, manure and wool. Ewes of this breed lamb thrice in 2 years and the sale of young lambs also provides a good source of income. Sheep milk is used in tea, or made into yoghurt and buttermilk. The coarse wool protects the animal from extreme temperatures and

Fig. 12.2. The Deccani sheep breed is ideally suited to the extreme temperatures of the Deccan Plateau and capable of long-distance migration – a necessary coping strategy during years of drought. Credit: ANTHRA.

weather patterns, which are typical of the semi-arid Deccan plateau.

The spinning and weaving of wool into coarse blankets and rugs are allied activities practised sometimes by shepherd groups themselves and sometimes by other communities. The wool is used to weave blankets called gongali/gonghadi/ kambali (a local blanket) and felted floor throws called jenn. These blankets have multipurpose uses and used to be very popular with shepherds and farmers in the region.

ANTHRA's work

Since 2004, ANTHRA has been involved in organizing the shepherds into village-level collectives and encouraging them to come together on a common platform to take collective action to address their problems. ANTHRA facilitates a process whereby information and technical support is shared amongst shepherd groups to enable them to discuss strategies to improve the health of their small ruminants, preserve and protect local animal genetic resources, improve fodder availability, address issues related to access to key natural resources such as water, grazing lands and pastures, and access services from the government veterinary health department. Far from being prescriptive, it encourages the community to identify their problems as well as work on possible solutions. A major component of the work is encouraging shepherding communities to conserve the local Deccani sheep breed, which plays a critical role in the livelihoods and agriculture production systems of the pastoralist and agro-pastoralists of the Deccan.

Why are we losing the Deccani?

Several factors have resulted in the rapid decrease of the breed in its traditional breeding tract. It began with the collapse of the coarse wool market as a consequence of large-scale dumping of merino wool from Australia and South America in the Indian market in the 1990s. People could now obtain apparel of finer wool at lower prices. Further, synthetic alternatives, which were finer and more durable, also became available in the Indian market and farmers who used to buy these coarse wool blankets every year shifted to more colourful and longer-lasting synthetic blankets. A large buyer of these coarse wool blankets at one time was the Indian army, who used to distribute these blankets to soldiers posted in cold mountainous areas.

However, again with increased global trade, cheaper synthetic blankets replaced the coarse Deccani blankets and the army stopped buying these blankets. With no market for the wool, even shearing of the wool became unprofitable for the shepherds. At the same time, there was an enormous increase in the demand for meat in the country as well as for export to other countries. The government, through the state animal husbandry departments, encouraged shepherds to replace their Deccani sheep with heavier non-wool sheep breeds. In Andhra Pradesh, shepherds were given loans to shift to the Red Nellore, in Maharashtra to the Madgyal breed and in Karnataka to the Yelugu breed. Unfortunately, the mutton breeds promoted by the state have been found to be more susceptible to diseases and less capable of coping with the stresses of migration. Being heavier, they require greater quantities of feed and fodder. The Madgyal has also been found more prone to diseases like orchitis in male breeding rams.

The Deccani is not a rare breed confined to a small geographical tract. In fact, its geographical spread is quite large (Fig. 12.3). However, it has a definite ecological, sociological and economic niche, and unless this niche exists, the breed is threatened.

One of the first steps therefore was to define the breed, its geographical spread, the specific niches that contributed to its survival and finally find simple strategies that could keep the breed alive.

Describing and defining the breed

Although considerable institutional research has been undertaken on the Deccani in

Fig. 12.3. Most shepherds – especially older shepherds – indicate that unless a favourable environment is created, they find it difficult to continue breeding the Deccani sheep. Credit: ANTHRA.

India, the studies were almost entirely conducted in a narrow geographical pocket of only one state, Maharashtra. The description of the breed within Indian research institutions was therefore incomplete. ANTHRA therefore undertook a focused 1-year study from March 2007 to March 2008 with select sheep flocks in the three states. The study aimed to document the breed as perceived and described by shepherds who breed the sheep as well as understand their traditional management and breeding practices. The different aspects researched included the traditional ways in which shepherds identify their animals, methods of selecting future breeding rams and ewes, and the care and management of animals in their flock. It also included monitoring and recording the growth and reproductive performance of the breed and its genetic variability. Parameters studied included the birth weight, body weight gain, lambing performance, etc. Wool was sampled and sent for testing. Grazing practices were also recorded.

The study revealed that shepherds have their own unique way of describing their breed and identifying their animals based on sex, ear size, body markings, wool colour, horn type and age. They have very specific parameters, based on which they select young lambs as future breeding rams, and ways of identifying the best ewes. Most

shepherds select the breeding ram from their own flock and prefer lambs that are born in the September–October lambing as abundant fodder post-monsoon favours good growth. Most lambing also occurs in this and shepherds have a larger group from which to select. Shepherds select the ram lamb as a future breeding ram when it is about 6–7 months old. The traits/characteristics observed by the shepherds for a future breeding ram include the weight and size of the ram lamb, its thighs, legs and rump, and its horns, which they prefer to be aligned close to either side of the face, pointing slightly outwards. In addition, they examine the ears, eyes, wool and overall appearance, they observe the mother's traits such as milking ability and udder size, and her ability to mother the lambs. Most of the shepherds prefer black ram lambs, as this conforms to the predominant colour of the Deccani flock.

They also believe that the black wool on the body of the sheep is important so that the sheep can withstand disease. When a breeding ram is selected from another flock, they look at the general appearance of the animal, wool character, legs, loin strength and width, and chest. After hearing the history of disease, they compare the details to the price being quoted before making a decision. When they select ewes, they look at the general appearance, conformation to the breed type, weight, size, and udder, and factors such as milk production, growth of lambs, season of lambing, disease history and mothering ability.

The quantitative data revealed that birth weights were higher in some districts compared with others and weight gain was directly linked to management practices where the addition of feed supplements enabled rapid weight gain in young lambs. This in turn was linked to the market and its demand, and shepherds were found to adjust strategies to be able to match the needs of the market. It was further observed that in all flocks there is a steady increase in body weight up to 2 years of age. Thereafter, the animal reaches its full body weight, and there is very gradual weight gain after this.

Wool is sheared twice a year; in Andhra Pradesh the wool is sheared in April and October, in Maharashtra it is done in February and August. Interestingly, in Andhra Pradesh the shepherds find that the wool yield, quality and smoothness is better in the second shearing in October, but in Maharashtra shepherds say that they obtain these qualities in the first shearing. Wool samples from the flocks were collected and analysed. Wool was finer in districts Narayankhed and Kolhapur where there was less mixing with other breeds such as the Madgyal and Red Nellore, which are primarily mutton breeds.

Deccani breed competition

To revive an interest in the breed it was decided to hold Deccani breed competitions annually. These were almost the only competitions held in the country for small ruminants. Most competitions in India are held for large ruminants or poultry. 'Best animals' in different age categories were identified. Also identified during these competitions were the best breeders of the Decanni. These competitions were also efforts to celebrate the possession of and invoke pride in the Deccani breed. The panel of judges included knowledgeable shepherds from other districts as well as officers and scientists from the government animal husbandry department and sometimes even research institutions such as the National Bureau of Animal Genetic Resources (NBAGR). Through this process, the officers and scientists get a chance to understand local realities and challenges faced by shepherds.

These events have proven to be a resounding success in multiple ways. Pastoralists themselves are extremely enthusiastic and a winning animal from their village is considered an enormous source of pride. Following the competition, many shepherds who owned Nellore or Madgyal breeding rams have come forward to change the rams and replace them with Deccani rams from other villages in their area. Shepherds in Andhra Pradesh and Maharashtra have begun to demand that Deccani rams be distributed through government programmes. The competitions have created a lot of interest about the local breed not only amongst shepherds but the wider community of farmers. The competitions also facilitate an understanding of ways in which shepherds traditionally select sheep. Most importantly, it has also evoked interest amongst the scientific community, to take a fresh look at the breed and the reasons they are kept.

Case Study 3: Bio-cultural Community Protocols to Support the Samburu Red Maasai Sheep and their Shepherds

By Jacob Wanyama, LIFE Network – Africa Region (www.pastoralpeoples.org)

Livestock keepers all over the world and especially in Africa depend heavily on livestock and their products to ensure their livelihoods. In Africa, 70% of the population practises livestock keeping not only for their survival but also for the survival of their future generation. This is because livestock keeping is not only the main livelihood option they have known since time immemorial, but also because it is more often than not the only viable option in the areas in which they live.

The practice of keeping livestock is commonly referred to as pastoralism, and is practised by many communities in Africa, but these are not the only populations that rely on livestock for livelihoods. Smallholder farmers who often find themselves in marginal areas also keep livestock to supplement their livelihood options. In Africa, livestock keepers occupy more than 70% of Africa's land mass and are found in areas otherwise not suitable for arable production. In fact, livestock keeping is the only economically and environmentally viable means of land use in these areas. It is the only efficient means of producing food and receiving some income at household level. Livestock also contribute a lot to the national economies of countries found in these regions. In Kenya, for example, it is estimated that livestock contributes about 10% of the Gross Domestic Product and accounts for over 30% of farm gate value of agricultural commodities (Government of Kenya, Office of Public Communication, 2007).

On a global scale, local livestock breeds significantly contribute to the efficient use of otherwise harsh environments and provide a valuable reservoir of gene pool, which can serve as an insurance against future challenges such as those brought about by climate change.

Pastoralists and smallholder farmers have kept domestic animals for centuries. To meet their many needs, local communities have domesticated and bred animals for different purposes, using specific selection criteria. These communities have in the process produced a wide variety of livestock types and breeds that are suited to the environment in which they are kept. In the process, these communities developed a wealth of knowledge on how to breed and care for these animals. Livestock keepers are therefore considered custodians of the world's livestock genetic diversity.

What has gone wrong?

Despite the fact that pastoralists and smallholder livestock keepers are guardians of much of the world's livestock diversity, their capacity for carrying out this work has been undermined to their detriment and to that of world food security. For decades, local or indigenous livestock breeds were regarded as inferior to the 'high-performing' breeds from the developed world. As a result, local livestock keepers are not given a chance or a choice to maintain their indigenous livestock breeds (Mathias *et al.*, 2005).

In a bid to achieve food security for ever increasing populations, developing countries have for years encouraged importation of 'high-producing' genetic materials mostly from the temperate countries to crossbreed with local breeds. Government and other development agencies developed and implemented programmes that often promoted replacement of indigenous breeds with exotic ones. The assumption was that this will inject the 'superior' genes into the local population and hence enhance food production. The resultant crossbreeds, however, in addition to being too expensive to be maintained by the resource-poor livestock keepers, have slowly eroded the region's animal genetic diversity. For example, it is now very rare to find a pure East African zebu in the so-called high potential areas. It is estimated that the livestock genetic pool especially for cattle in high potential areas of Kenya has reduced by 50%. To date, even the 'road-side grazing system' practised by the landless is characterized by crossbred cattle (Rege, 2001).

This was happening under the backdrop of the fact that livestock keepers are increasingly losing their production resources as grazing lands and services, and are increasingly facing the effects of climate in the form of drought. As a result, many livestock keepers are giving up their way of life and seeking alternative livelihoods, as they can no longer feed their families on livestock keeping only. This situation is being worsened by the increasing economic pressures on livestock keepers' households forcing them to seek more economically profitable enterprises including changing their livestock herds to the so-called more economically valuable breeds. All these changes are putting many indigenous livestock breeds in danger of becoming extinct. For example, in Kenya up until the 1970s, there were several million Red Maasai sheep (Rice, 2008; Taberlet, 2008).

Despite the fact that this sheep is more resistant to environmental pressures such as intestinal parasites, it has been indiscriminately crossbred with the woollier and much larger imported sheep breeds in response to market pressure. This is nearly driving this sheep breed out of existence. Another indigenous livestock breed under economic pressure is the Longhorn Ankole cattle. This is a dairy breed developed by

Bahima pastoralists in Uganda and Rwanda, which is highly valued locally for its tasty milk with high fat content. However, indiscriminate crossbreeding with exotic dairy cattle is rapidly diminishing the number of purebred Ankole. This dire situation has been acknowledged internationally, where it has been recognized that many domestic animal breeds are threatened with extinction. An FAO study estimated that every month more breeds of domesticated animal disappear from the face of the earth (FAO, 2007).

Value of local breeds more evident

Despite this apparent abandonment of indigenous livestock breeds by livestock keepers and lack of recognition of the same by development agencies, the value of local livestock breeds and their advantages over high-performance breeds is increasingly becoming evident (Fig. 12.4). More and more scientific studies are indicating that the performance of indigenous breeds is equal to or even better than that of improved or crossbred animals in similar environments. For example, in Kenya, the Dorper is a sheep breed originally from South Africa introduced in Kenya to crossbreed with local sheep breeds to increase production. An experiment to compare the productive performance of the Red Maasai

Fig. 12.4. The Red Masaai sheep has unique characteristics, like disease resistance and high fertility. Credit: Jacob Wanyama.

and the Dorper sheep breeds under contrasting environments in Kenya showed that: 'the indigenous Red Maasai sheep were 3 to 5 times more productive and efficient than Dorper sheep in the humid coastal environment. The Dorper were only slightly more productive than the Red Maasai in the semi-arid environment, with no significant differences in flock efficiency between the breeds' (Okeyo and Baker, 2005).

In addition, certain livestock breeds have been scientifically proven to have unique genetic characteristics such as disease resistance and high fecundity that could be useful for future development of the livestock industry to address world hunger and poverty. However, while this increased importance of these unique characteristics of indigenous livestock may lead to more efforts towards the conservation of these unique breeds, it has an inherent danger of also leading to the misappropriation of these unique characteristics for commercial interest at the expense of the livestock keepers.

What is the remedy?

In order to remedy the situation, livestock breeders and planners need to develop and put in place programmes and policies that promote the sustainable management of these breeds by incorporating indigenous breeding practices into governmental breeding programmes, increasing the value of indigenous breeds, motivating indigenous livestock breeders to continue keeping these breeds, and perhaps most important including indigenous keepers as equal partners in conservation efforts. LIFE Africa together with its local partners is attempting to address some of such issues.

LIFE stands for Local Livestock for Empowerment of Rural People. It is an action research and advocacy network that aims to secure and improve the livelihoods of livestock keepers by promoting sustainable use and conservation of local livestock breeds. To do this, the LIFE Network is creating awareness of the value of local livestock breeds through:

- Promoting community-based breed conservation initiatives through motivational activities such as local breeds shows, giving pure local breeds conservation awards and development of niche markets;
- Conducting participatory documentation of local breeds and development of Bio-cultural Protocols;
- Conducting training and capacity building of livestock keepers and development workers on participatory documentation of indigenous livestock breeds;
- Conducting participatory research and publication on topical issues regarding conservation of indigenous livestock breeds;
- Advocating and influencing policies, through constructive dialogue, to recognize the vital role played by the local livestock keepers in the conservation and sustainable utilization of indigenous livestock breeds, and the need to support them to continue doing so and obtain a share in any benefits that may accrue from any commercial exploitation of the unique characteristics of these breeds.

The LIFE Network is an open membership network with members from NGOs, herders' associations, scientists, development workers, volunteers and individual supporters. The network has an international secretariat based at the League for Pastoral Peoples in Germany and regional coordination secretariats in Africa, based in Kenya, and in Asia, based in India.

The case of the Red Maasai sheep

The Red Maasai sheep is among the indigenous livestock breeds that have been bred and conserved by local livestock keepers for centuries. It is an indigenous sheep of Kenya and Tanzania, which belong in the class of fat-tailed or fat-ramped sheep (Rege *et al.*, 1996). They are recognized for their ability to walk long distances in search of pasture, adaptability to the harsh environmental conditions (Owen *et al.*, 2005) and have unique traits for resistance to gastrointestinal nematodes (Baker *et al.*, 2002). However, as a result of indiscriminate crossbreeding with exotic breeds especially with the Dorper and the Merino sheep, this sheep is faced with a serious threat of extinction.

Generally, the Red Maasai sheep is identified with the Maa-speaking communities of Kenya and Tanzania, who are considered to have created and conserved this sheep for centuries and used it for its fat and meat. These communities comprise mainly of the Maasai of southern Kenya and northern Tanzania, and the Samburu of northern Kenya (Fig. 12.5). However, the sheep can also be found among the neighbouring tribes such as the Nandi and the Bukusu in Kenya, the Gogo in northern and central Tanzania and the Karamoja in the drier parts of Uganda.

The sheep's habitats are found at an altitude range of 500–1500 m, in a semi-arid climate with bimodal rainfall, and in pastoral or agro-pastoral production systems. Scientists believe that the Red Maasai sheep belongs to ancestral fat-tailed sheep that entered Africa at various occasions through both the straits of Suez and Bab el Mandeb at the beginning of the second millennium. The group that entered Bab el Mandeb later extended from Ethiopia into the lake regions of Uganda, Kenya and Tanzania. It is believed that the relevance of the fat deposit in the tail to pastoral communities as a source of energy-rich food contributed to the extensive replacement of the original thin-tailed sheep by the fat-tailed types.

The Red Maasai sheep initially attracted the attention of scientists because of its large population, wide distribution, uniformity of body colour and hardiness (Kiriro, 1994). However, more recently, scientists have confirmed its unique genetic capability to cope with parasitism better than other sheep breeds. Currently, there are reports that major research institutions and large sheep-rearing countries are looking into these possibilities (Sieper, 2007). This raises concerns among the indigenous Red Maasai sheep breeders of possible misappropriation of this sheep's genetic resource at their expense. At the same time, the survival of both the Red Maasai sheep and its indigenous breeders is threatened by various factors. These include lack of support and services from development agencies, increased droughts decimating livestock numbers, inappropriate government policies that promote settlement of pastoralists, and indiscriminate crossbreeding of this sheep with the so-called high-value breeds.

Fig. 12.5. Samburu men with a herd of Red Maasai sheep rams. Credit: Jacob Wanyama.

Bio-cultural Community Protocol

In 2010, LIFE Network Africa in collaboration with Natural Justice assisted a group of the Samburu Community to develop a Bio-cultural Community Protocol (BCP) on the Red Maasai sheep. The objective was to articulate the integral role played by the Red Maasai sheep as well as the other indigenous livestock breeds in their culture and livelihoods. The protocol aimed at enabling communities to make a case for *in situ* (rather than *ex situ*) conservation of their breeds under Paragraph 8j of the international Convention on Biodiversity (CBD). Moreover, the process of developing a BCP was aimed at empowering the community by helping them to think through various opportunities that may be inherent in their animal genetic resources.

The development of Samburu Bio-cultural Protocol was a follow-up of a similar process, which had been spearheaded by the LIFE Network Asia and Natural Justice among the Raika camel herders community in Rajasthan, in northern India (Köhler-Rollefson, 2010). The successful completion of the Raika protocol clearly demonstrated the potential of this tool for strengthening communities who steward indigenous breeds.

The process of developing the Samburu Bio-cultural Protocol started off with a participatory documentation of local knowledge on the Red Maasai sheep by the Samburu Community in Samburu District of Kenya. This was carried out to facilitate the Red Maasai sheep keepers to develop their own views on the rearing of this sheep and package the information generated into an advocacy piece of work. Participatory documentation of indigenous livestock breeds methodology, which has been pioneered by the LIFE Network, enables livestock keepers together with scientists to document and record their own breeds; it documents the local perceptions and use of the particular breed. The data collected was analysed and feedback was given to the respective communities.

In the case of the Samburu Community, the information so generated was also used to develop a draft Bio-cultural Protocol with support by lawyers from Natural Justice. The draft Protocol was used to obtain feedback in a workshop, during which representatives of the community discussed, corrected and endorsed the content of the protocol. The endorsed draft was then finalized and printed into a document written in two languages – Samburu and English. A summary of the results of the participatory documentation is shown in Box 12.2.

Box 12.2. Information from Samburu herders on their Red Maasai sheep.

---

1. Socio-cultural association of the Red Maasai sheep to the Samburu Community
The Red Maasai sheep is part and parcel of the Samburu society and culture. It is very important during ceremonial rites of passage such as marriage. During these ceremonies, the groom is required to present a pure Red Maasai sheep to his mother-in-law, as part of the process of cementing the new relationship. During circumcision, the circumcised boy is required to wear a sheep's skin before and after circumcision. He is also required to sit on the sheep's skin when being circumcised. At the time of graduation from boyhood to manhood, a Red Maasai sheep is slaughtered for the transition ceremony. The sheep is also slaughtered during the birth of a new baby to celebrate this occasion as well as provide meat for the mother during the postpartum period. The Red Maasai sheep is also an important source of traditional medicines. The sheep's fat is believed to cure snakebites and measles. After the death of an old person, the sheep's fat is smeared on the deceased's mouth as a sign of last respect and wishing him or her good life in the next world.

2. Breeding practices
Most respondents indicated that the best young Red Maasai rams are usually selected and then the rest is castrated. The reason for this is for one to have a healthy flock. Most shepherds of the Red Maasai sheep buy their breeding stock either from the nearby government farm or National Youth Service Farm.

Continued

Box 12.2. Continued.

Apart from buying and selecting breeding rams within one's stock, one can also exchange a female sheep with a neighbour's young ram. Sometimes, there is also borrowing of a breeding ram from a neighbour, friend or relative. Herders can also take a sheep in heat to a neighbour's flock for mating. Breeding usually takes place during rainy season or at a time when the sheep are healthy and in a good body condition.

The breeding is controlled but not fully. When there is an outbreak of any disease, breeding is not allowed by the community until the outbreak is under control. This is also true during hard times such as famine. To avoid mating, the sheep are given skin aprons, and sometimes the rams are separated from the rest of the flock. However, it is very hard to control breeding or mating when the flocks are out for grazing as the rams may get loose and mate freely within and outside the flock.

3. Local perception about the origin of the Red Maasai sheep
Most of the young respondents indicated that they did not know the progeny of the Red Maasai sheep and were not aware from where it first came. However, the older people said the sheep originated from the Njemps of Baringo, others said it came from the Maasai Community in the Rift Valley, while a small percentage say the sheep came from 'Lchogi', which is a place in the neighbouring district of Laikipia district.

The community refers to the sheep as 'Nker Nanyokie' meaning red sheep. They said that despite the fact that the sheep is referred to as red sheep, black coloured sheep or other colours can also be found among the flock but they are still referred to as red sheep.

4. Production and productivity
Most who responded said the Red Maasai sheep is dominantly found in Ldonyo lo Loroki (i.e. Loroki and Kirisia divisions of Samburu District), a plateau area of the district. They noted that in places where Red Maasai sheep are found, the rain is enough to support small-scale farming. Hence, the livelihood in this zone is agro-pastoralism that covers 15.4% of the district. Recently, some people from outside the area have started moving in and fencing their land, restricting access to grazing. The sheep do not follow a seasonal forage calendar, as they are resistant to drought. The Red Maasai sheep like many other local multipurpose breeds offer products like milk, meat and fat. In contrast to other communities, Samburu herders also drink its blood. The sheep give birth twice a year starting from 1 or 1.5 years of age; they give birth up to 10–12 times in a life time. Sometimes the sheep produce twins.

5. Population trends and conservation
The study was not able to estimate an exact population of the Red Maasai sheep. However, field observation and interaction with herders indicated that the pure type of sheep is disappearing at a very fast rate. Most of the flocks visited consisted of less than half Red Maasai sheep. The remainders were exotic purebreds and crosses, mostly with the Dorper sheep introduced during the early 1990s. Initially, the local people preferred the Dorper breed because of its heavy body weight, which fetched higher market prices, and also because they were told that the Dorper sheep were superior to their own breed, but later the local breeders realized that the Dorper sheep and its crosses with the Red Maasai sheep had serious disadvantages. For example, this breed could not resist drought and diseases in the area.

In general, the respondents were astonished that the interviewers were stressing the advantages of the Red Maasai sheep after other previous extension agents had told them that these sheep were useless. However, most of them strongly felt that they should not lose the breed and urged for outside help in order to sustain the sheep. The communities acknowledge the value of the Red Maasai sheep. For example, a Red Maasai breeder, Mr Lerija Ledanako, said that the Red Maasai sheep unlike any other can trace their way home when they get lost while grazing, which reduces the chance of their being preyed upon by wild animals.

What were results of this process?

In addition, the process of documenting and developing the Bio-cultural Protocol has had a tremendous impact on the perception of the Samburu indigenous Red Maasai sheep breeders. The Bio-cultural Protocol has enabled these communities to flag this unique breed of sheep as their creation and property under their own custodianship.

The first step towards this was during the feedback workshop of the draft Bio-cultural Protocol, which was held with the wider group of the Red Maasai sheep breeders including those that participated in the process of data collection and development of the protocol. During this feedback session, the communities, in addition to discussing, correcting and endorsing the content of the protocol, were able to put forward what they felt was the best way forward for the use of the protocol and their indigenous livestock breeds. They expressed the need for the Bio-cultural Protocol to be used as a learning tool for their young members of the community, as well as a tool to inform the world of the community's existence and its contribution to global biodiversity. They also agreed to organize themselves at community level and scale-up the coverage of the protocol to other indigenous groups that keep the Red Maasai sheep.

An official launching of the protocol was organized in Maralal on 28 May 2010 and officiated by Government representatives from Kenya's Ministry of Livestock and Fisheries. During this event, the Red Maasai sheep breeders were able to air their views regarding the government role in providing services to them, and the need to be supported to continue playing their central role in conserving indigenous animal genetic resources.

What is the way forward?

As a result of this process and in response to the feedback from the Samburu community that participated in this process, the LIFE Network in collaboration with the Samburu community and the relevant formal institutions will continue to facilitate this process. Some of the activities planned include:

- Promoting the sharing and endorsement of the protocol by the rest of the Red Maasai sheep breeders in the region (including the Maasai of Kenya and Tanzania);
- Using the protocol as a tool to creating awareness among the policy makers, scientists and development workers regarding the plight of the Red Maasai sheep and the communities that keep them;
- Advancing the issue of livestock keepers' rights at national and regional level with the specific purpose of influencing and advocating for scientists and governments to recognize and respect the role of livestock keepers as creators and custodians of indigenous livestock breeds and working with them not only as givers of information and materials but as decision makers and equal partners.

References and Further Reading

Baker, R.L., Mugambi, J.M., Audho, J.O., Carles, A.B. and Thorpe, W. (2002) Comparison of Red Maasai and Dorper sheep for resistance to gastro-intestinal nematode parasites, productivity and efficiency in a sub-humid and a semi-arid environment in Kenya. *Proceedings of the Seventh World Congress on Genetics Applied to Livestock Production*, Montpellier, communication 13-10.

FAO (2007a) *The Global Action Plan for Animal Genetic Resources and the Interlaken Declaration*. FAO, Rome.

Kiriro, P.M. (1994) *Estimate of Genetic and Phenotypic Parameters for the Dorper, Red Maasai and their Crosses*. ILRI, Nairobi.

Köhler-Rollefson, I. (2010) *Bio-cultural Community Protocols for Livestock Keepers*. League for Pastoral Peoples and Endogenous Livestock Development (LPP) and LIFE Network.

Mathias, E., Köhler-Rollefson, I., Geerlings, E. and van't Hooft, K. (2005) Endogenous livestock development – can it help the poor? *Tropentag 2005, Conference on International Agricultural Research for Development*, Stuttgart–Hohenheim, 11–13 October.

Okeyo, A.M. and Baker, R.L. (2005) *Methodological Illustration of Genotype × Environment Interaction (G×E) Phenomenon and its Implications: A Comparative Productivity Performance Study on Red Maasai and Dorper Sheep Breeds under Contrasting Environments in Kenya*. ILRI, Nairobi.

Owen, E., Kitalyi, A., Jayasuriya, N. and Smith, T. (2005) *Livestock and Wealth Creation. Improving the Husbandry of Animals Kept by Resource-poor People in Developing Countries*. Nottingham University Press, Nottingham.

Rege, J.E.O. (2001) Defining livestock breeds in the context of community-based management of farm animal genetic resources., Key note speech in Workshop on Community based management of animal genetic resources: A Tool for Rural Development and Food Security, Mbabane, Swaziland, 7–11 May.

Rege, J.E.O., Yapi-Gnaore, C.V. and Tawah, C.L. (1996) *2nd All Africa Conference on Animal Agriculture*, Pretoria, 1–4 April.

Rice, A. (2008) A Dying Breed. *New York Times* (www.nytimes.com; 27 January 2008).

Sieper, K. (2007) Red Maasai sheep helping Australia. ABC Online (www.abc.net.au; 11 December 2007).

Taberlet, P. (2008) Are cattle, sheep, and goats endangered species? *Molecular Ecology* 17, 275–284.

Case Study 4: The Heifer Methodology of Supporting Sustainable Livelihood in Kenya – The Story of Mr Laban Kipkernoi Talam

By Reuben Koech, Heifer International Kenya (www.heifer.org)

Mr Laban Kipkernoi Talam is a youth by Kenyan standards. However, he is married to Miriam Talam and manages his family farm in Kipchomber Village, Kabiyet Division, Nandi County. The family has two children, Pacificas Cherop, 9 and Eugene Kipchirchir, 7. Both children are enrolled in private school in Standards 3 and 1. Laban's family is also taking care of an orphan, Kevin Kipchirchir, 7, who is also in Standard 3.

Laban has already had a full life of struggle and determination before and after he was married. Even though Laban had a 1-week stint in high school, he is by Kenyan standards a lowly educated person because he left high school because of a lack of school fees. To his credit, this did not kill his spirit, as he immediately started farming from his inherited 2-acre piece of land. He started horticulture farming with a focus on tomatoes, kale and aubergine.

In 2002, Laban 'borrowed' a dairy cow from his relative to help him take care of the nutrition of his family and to provide some income. To meet his family's ever-growing needs, he decided to venture into a brick-making business. He also expanded his horticulture farming business to include capsicum and other local vegetables. He used to rent 0.2 acres of land for the gardens and was paying Ksh. 1000 (~US$13.30)/month. In 2005, after seeing the benefits of being in a successful group in the area, he started a group of his own with 16 members, called the Silanga Youth Group. The group was mainly involved in merry-go-round activities, such as horticulture and tree planting.

In 2005, the group applied for funds from the Government of Kenya Youth Enterprise Development Fund (YEDF). The group was, however, not successful because they had no bank account.

In 2007, they tried again for the same funds. This time, they were successful, as the group had established a bank account. They were given Ksh. 46,600 (~US$621.30). As a group, they used the money to expand their horticulture farming business from 1 to 3 acres, with each member having a portion of between 0.2 and 0.3 acres. They cleared the loan from YEDF in 2008 and used the profits that they had accumulated in their account to buy a dairy heifer for the group.

Up to this point, Laban's family had the following sources of income before involvement in their community group:

- Horticulture farming business, which generated income of Ksh. 17–20,000 every 3 months. Laban sold tomatoes from house to house in area towns.
- Dairy milk sales from his 'borrowed' dairy cow earned an average of Ksh. 1200/month. The price per litre of milk sold to brokers was between Ksh. 8 and 11/l.
- Black market business generated Ksh. 15,000/year.
- Laban started planting 300 eucalyptus tree seedlings on his farm. He has managed to use the trees to build the zero-grazing unit for his livestock, to fence his land and to sell as firewood.

Then, a new programme was introduced into the area. The East Africa Dairy Development (EADD) programme began working with dairy farmers, such as Laban, in Kabiyet in August 2008. EADD started mobilizing farmers through training meetings. Laban and his group received training through demonstrations on animal husbandry, fodder management, soil conservation and record keeping. The group also started buying shares from the newly registered Kabiyet Dairies Ltd. The demonstrations that were held at Laban's farm included planting of Elba Rhodes, *Calliandra*, Kakamega, sweet potatoes, lucerne and *Desmodium*. As the training continued, Laban was selected by the EADD team for a Training of Trainers course. He started using his acquired skills to train his group and other Dairy Management Groups that were developed by the EADD (Laban's family are pictured in Fig. 12.6). Laban now trains,

Fig. 12.6. Laban and his family on their farm.

on average, 60 farmers per month from in and out of the county on animal husbandry.

From the EADD and Government of Kenya extension staff training, Laban has managed to change drastically his dairy cow management practices, particularly on on-farm feed formulation. This has greatly boosted his milk production from an average of 5 to 20 l/day. It has also drastically reduced his costs of buying expensive feed supplements from the agrovet shops.

When the Kabiyet Dairies Ltd., opened its doors, Laban was one of the early milk suppliers. From his 'borrowed' cow, he could manage only 5 l of milk daily to the dairy. To facilitate payments of milk proceeds, Laban opened his own first ever bank account with Equity Bank. With regular milk payments to the bank, he has become creditworthy and has been given a loan of Ksh. 20,000 (US$267) to buy his own dairy cow, guaranteed by the dairy plant and a 3-month bank statement. He had also managed to save Ksh. 10,000 from his milk proceeds. He now has Ksh.30,000, which he used to buy an in-calf heifer, which calved 1 month later. With the new dairy cow, milk supply to the cooler increased from an average of 5.5 to 20 l, and with the new good price of milk at Ksh. 25/l offered by Kabiyet Dairies, Ltd., he is earning over Ksh. 500/day,

meaning Ksh. 15,000/month. This is an increase of over 1150% compared with what he was earning before. He also manages to sell 3 l of evening milk to neighbours at Ksh. 20/l, which brings in an additional Ksh. 60/day or Ksh. 1800/month for his wife, Miriam, for her own household needs.

With the increased demand for fodder by the dairy cows, Laban buys Napier grass from neighbours to make silage for the dry season, as his 2-acre piece of land could not provide enough. Laban has also started using AI services to breed his cow. In addition, with milk proceeds savings, he managed to buy the 'borrowed' cow from his relative for Ksh. 22,000 and became a proud owner of two dairy cows.

Human capital asset change

Laban and his wife Miriam have greatly benefitted from the trainings received from EADD. As a result, Miriam can easily direct and manage dairy animals with ease as Laban goes about his other business, including the Training of Trainers.

Laban's family has managed to move his children, including the orphan, from the public schools to private schools with better schooling and academic facilities. Laban, in his own words says, 'We can now afford to have meat with our meals anytime we want, unlike before. We used to eat meat only during Christmas holidays'.

Financial capital asset change

Before EADD, Laban's family never had any savings. It was an almost hand-to-mouth family struggle. Now, he can boast of Ksh. 12,000 at the end of every year, despite having soaring expenditures. He now has 12 sources of income for his family. He has also bought shares in the Kabiyet Dairies, Inc., worth Ksh. 5000, from which he hopes to get dividends soon.

Laban is an astute young man who uses every opportunity he comes across for his benefit. According to his neighbours, he is a brave and honest man out to try all skills that come his way. He is an early adopter who would stop at nothing but keeps trying new things until they work out. From the trainings he received, Laban has managed to develop and strengthen his income sources in even additional ways:

- Using milk proceeds, he has invested in poultry and rabbit farming, enabling the family to earn additional income;
- He expanded the horticulture business;
- Laban supplies homemade Robyn hand washer pumps, selling on average two per month;
- He uses his fodder chopper/pulverizer to chop feed for other farmers;
- He makes silage for other farmers;
- He uses his motorbike to attend distant training, to transport milk to the dairy collection centre, to haul fodder and supplements – all to avoid higher transportation charges;
- Laban charges or seeks exchanges with other farmers for new training sessions;
- He opened an agency for *Calliandra* seeds, sweet potato vines and several sorghum varieties;
- He designs and lays out new dairy zero-grazing units.

Physical capital asset change

From milk proceeds, Laban can boast of having a zero-grazing unit, a biogas plant, a pulverizer (Fig. 12.7), a motorbike, wheelbarrow, a shallow water well and improvements on the house of an iron roof replacing grass thatch. He makes his own on-farm feed supplements from sunflower,

Fig. 12.7. Pulverizer or feed chopper for dairy rations.

sorghum, Columbus grass, Bhoma Rhodes, *Sesbania*, mulberry, *Desmodium*, sugarcane and Napier grass, all grown in his 2-acre farm (Fig. 12.8). He also buys some fodder from other farmers to supplement what his farm cannot produce.

Natural capital asset change

Laban can also now boast of having two dairy cows and one beautiful AI heifer calf. The fear of the 'borrowed' cow being repossessed at any time by the owner is now over.

He has also used his milk proceeds to acquire fertilizer, maize seeds and to use hired labour to help with the 0.2-acre garden. The land tilling used to be done by him and his wife because of lack of funds.

Social capital asset change

Besides Laban training other farmers through practical demonstrations, he is now a common figure on the local radio shows. He is invited on a regular basis by pharmaceutical companies who want to use his knowledge and experience to train and challenge other dairy farmers to improve

their income from farming. He has received numerous requests for his secret of success.

Does he inspire other young farmers like himself to do the same? Laban says that he is currently aware of 10 youths who seek information from him. One challenge in the area is that older men do not always easily release parcels of land to their sons and daughters while they are still young and energetic. A number of families deny this passing along to their children while they are young and able to till the soil. 'I want to use my influence now to change this behaviour forever!' says Laban.

Laban is also being called on for 'harambees' or village fundraising to help the poor. He normally donates Ksh. 500 per harambee and uses the opportunity to challenge youth and others to practise good farm business.

Besides his original group, he is also a member of another group called the 'Tepsainai Youth Group' and is involved in the establishment of tree nurseries, earning his family additional income.

Political capital asset change

Laban does not like politics. He has been urged on several occasions to run for a councillor position but has declined. However,

Fig. 12.8. Laban makes his own on-farm feed supplements from sunflower, sorghum, Columbus grass, Bhoma Rhodes, *Sesbania*, mulberry, *Desmodium*, sugarcane and Napier grass.

there is one thing that he can do and that is to influence the community to vote for the best Member of Parliament.

Future plans

- To have his children attend the university;
- To expand the farm to 5 acres to enhance production and income;
- To acquire a better producing dairy cow – one that will yield 28–30 l/day;
- To build a permanent main house;
- To explore other business ventures;
- To strengthen his charity work through harambees and to help HIV/AIDS patients.

For Laban and his family, training and good management has provided the opportunity for marketing that has increased the family income and resulted in a better way of living.

Case Study 5: Ethno-veterinary Medicine in a Master's Degree Programme on Tropical Agricultural Production

By Raúl Perezgrovas and Guadalupe Rodríguez, Instituto de Estudios Indígenas-UNACH, México (www. iei.unach.mx)

Around the globe, traditional animal health care approaches are experiencing a comeback, triggered by an increasing recognition of their value and the need to reduce the heavy use of chemicals in veterinary medicine and agriculture. In 1986, McCorkle (1986) coined the term 'ethno-veterinary medicine' for such approaches and – together with other scientists and development practitioners – advocated for their enhanced use in development. Since then, the number of studies, projects, documents and theses on ethnoveterinary medicine has been steadily increasing (Mathias-Mundy and McCorkle 1989; Martin *et al.*, 2001). Another comprehensive book on *Ethnoveterinary Botanical Medicine* (Katerere and Luseba) was published in April 2010.

However, veterinary faculties have been slow in picking up on this trend, offering information on medicinal plants at best, but ethno-veterinary medicine has much more to offer. A systematic integration of ethnoveterinary practices and information into veterinary curricula and education would not only widen the spectrum of prevention and treatment choices, it could also deepen the understanding of and respect for health care approaches of communities and make veterinary services in marginal areas more appropriate to the needs of livestock keepers. Plus, it would cater to the increasing demand among clients for alternative health care approaches.

University of Chiapas, Mexico

A particular graduate programme in tropical agricultural production has been recently presented by the University of Chiapas, in Southern Mexico. This Master's programme is unique in many ways, mainly because it is offered at four different campuses, taking advantage of the diversity of the staff, and benefiting from the different facilities at each university location. In this programme, there is a strong technical input provided by two Schools of Agronomy (one oriented to the tropical production systems and the other to the commercial production of maize and beans in more temperate climate) and the School of Veterinary Medicine; the social and cultural aspects of the agricultural production rely on the expertise acquired over the years by the staff at the Institute for Indigenous Peoples' Studies.

Another characteristic of this innovative graduate programme is that there are only three mandatory courses: Statistical Analysis, Tropical Agroecology and Socioeconomic Diagnosis of Agricultural Production. The rest of the curriculum is designed with the students by a group of academic counsellors, choosing from a diverse list of 38 elective courses taught at the different university grounds.

One of the interesting features of the graduate programme in Tropical Agricultural Production is the flexibility of the curriculum and the mobility of the students. This means that a student can go to a different university (national or international) if one of their courses or any other academic activity (e.g. workshops, seminars) is recommended by the tutors.

Ethno-veterinary Medicine and Ethno-animal Husbandry

It is important to mention that two of the elective courses in the graduate programme are related to ethno-veterinary medicine and traditional agricultural systems. The course on Ethno-veterinary and Ethno-animal Husbandry may be one of the few graduate courses offered on the ethno-veterinary discipline, not only in Mexico but also worldwide. It involves theoretical work that requires the review of the classical publications that began the discipline, with articles and books by Constance McCorkle, Evelyn Mathias-Mundy and Marina Martin.

The course on Ethno-veterinary and Ethno-animal Husbandry has a strong content on the methodological aspects of the study of animal production systems, utilizing the tools provided by the social sciences, and it suggests the analysis and discussion of the empirical information gathered in the field based on a multi-disciplinary approach. A great component of this course requires extensive fieldwork among the indigenous communities that continue living in Chiapas.

Management of Animal Genetic Resources

There exists an additional elective course in the graduate programme on Management of Animal Genetic Resources, which promotes the study, conservation and promotion of the local breeds of domestic animals, along with the study and conservation of the traditional systems of animal husbandry. This course has a unique orientation on rural development based on the traditional livestock systems. The different ethnic groups found nowadays in Chiapas, not less than nine, make it possible to look into ancient and efficient systems of animal production, and also bring about the possibility of appreciat-ing a good number of local animal breeds of the different species.

The graduate programme on Tropical Agricultural Production is in its fourth year, and already has obtained the second classification of 'Program in Development' from the federal authorities, in contrast to the initial level of 'Program under Formation' and will try to acquire the ultimate rank of 'Consolidated Program' in the near future.

New undergraduate programme

Making use of this academic experience at the graduate level, the staff at the Institute for Indigenous Peoples' Studies has been invited to collaborate in the design of a new course on ethno-veterinary medicine, which will be offered within the undergraduate programme on 'Sustainable Development' by the Intercultural University of Chiapas. This course will have as its core the practical aspects of information gathering with local experts, collection and preservation of botanic species of ethno-veterinary value, workshops for the preparation of herbal remedies, adding value to products by controlled quality processing, and proper marketing.

References

Mathias-Mundy, E. and McCorkle, C.M. (1989) *Ethnoveterinary Medicine: An Annotated Bibliography. Bibliographies in Technology and Social Change No. 6.* Iowa State University, Ames, Iowa.

Martin, M., Mathias, E. and McCorkle, C.M. (2001) *Ethnoveterinary Medicine. An Annotated Bibliography of Community Animal Healthcare.* ITDG Publishing, London.

Katerere, D.R. and Luseba, D. (eds). (2010) *Ethnoveterinary Botanical Medicine. Herbal Medicines for Animal Health.* CRC Press, Boca Raton, Florida.

Case Study 6: The Dairy Development Trap: How Developing Countries can Learn from the Experiences of Dutch Dairy Farming

By Katrien van't Hooft, Dutch Farm Experience, The Netherlands (www. Dutchfarmexperience.com)

History of Dutch dairy farming

Intensification of dairy farming

Only 50 years ago, agriculture in The Netherlands was quite similar to agriculture in many other (developing) countries today: large numbers of family farms that combine low-input crop production with various species of livestock for milk, meat, manure, traction and cultural manifestations. Until the early 1960s, milking was done by hand, carts were pulled by horses and fodder was dried as hay for the winter period, when the cattle stayed inside in rope-tied barns.

Since the 1960s, dairy farming in The Netherlands has gone through a metamorphosis. The average number of cattle per farm has increased sevenfold: from nine to 66 animals. Modern free roaming farms today can even keep up to 1000 animals. At the same time, one man in 2007 produces 17 times the amount of milk that one man in 1960 produced; the number of dairy farms has decreased by 85%: from 180,000 farms in 1960 to 21,300 in 2007 (Table 12.1).

What does all of this mean besides a major increase in quantity of milk produced on each farm? How was this achieved? In what way were farmers supported? Does

this imply that farmers in 2007 earned 17 times more than their fathers in 1960? What hurdles were met on the way? What are the trends today? And what can be learned from all this for other countries that want to improve and modernize their dairy system?

This case study presents the history of dairy farming in The Netherlands between 1960 and 2010. It will first highlight the conducive policies that supported the transformation in dairy farming between the 1950s and 1970s. Then the side-effects that started to appear from the 1970s are mentioned, followed by the policies put in place to reduce these consequences. Finally, the latest policies and trends will be mentioned, and the way local knowledge and farmer-driven initiative have again gained importance. I will conclude with the lessons that were learned on the way, which may be of use for dairy development in other countries.

Dairy policies between 1950 and 1970

After suffering a lack of food during the Second World War (1940–1945), the agricultural policies in The Netherlands were aimed at 'no more hunger': increased production of cheap food that would guarantee an income for the farmers. EU agricultural policies in the same direction started in 1957. This implied major government investments with the strategic focus of maximization of food production: obtaining the highest possible yields per hectare and kilogram of milk per animal per year through specialization, mechanization, intensification and scale enlargement.

As a result, while in 1960 the average Dutch dairy cow would produce 4200 kg of

Table 12.1. Dairy development in The Netherlands, between 1960 and 2007 (Ham *et al.*, 2010).

	1960	1975	1985	1995	2000	2005	2007
Dairy farms (×1000)	180	91.5	58	37.5	29.5	23.5	21.3
Total milk production (×1000)	6,721	10,286	12,525	11,280	11,155	10,827	11,134
Dairy cows (×1000)	1,628	2,218	2,367	1,708	1,504	1,433	1,413
# of dairy cows per farm	9	24	41	45.5	51	61	66
Milk production/farm (×1000)	37	112.5	216	301	379	460	522
Milk production (kg/cow/year)	4,200	4,650	5,300	6,610	7,420	7,550	7,880
Milk production (kg/ha/year)	5,500	8,864	12,512	12,018	12,340	12,560	12,980
Labour productivity (kg milk/h)	8	37	72	89	108	128	141

milk per year, in 2007 this had nearly doubled to about 7880 kg. Average milk production per farm per year has increased 14-fold: from 37,000 in 1960 to 522,000 kg of milk in 2007. This phenomenal growth was a result of successful technology development aiming at highest milk yields per animal per year. It was enhanced by effective research–extension–farmer interaction and easy access to credit. The market was protected by guaranteeing fixed prices and other measures of active government support to the agricultural sector (Box 12.3).

Blanket recommendations

Based on this general model of agricultural development, 'blanket' recommendations that applied to all farms in all regions and with all soil types were designed. Extension services, farmer education, research and agri-business (supported by the government) followed these recommendations.

The availability of high-quality roughage, supplemented with high levels of protein-rich concentrates, made it possible to exploit fully the improved genetic potential of the (predominantly) Holstein Friesian dairy cows, but also, the low prices of the high-quality fertilizers and concentrates were essential in achieving high milk production. Artificial insemination and effective breeding policies increased the potential milk yield of dairy animals to levels that our grandfathers did not even dream of.

In the process of specialization, animal and crop farming became completely divided, and milk production became divided from meat production. Increasing fertilizer application levels to the grasslands boosted grass yields and fodder production. Improved fodder conservation techniques and the introduction of fodder maize boosted milk production (Fig. 12.9). The obligation to collect the milk in milk tanks rather than in smaller containers also stimulated the shift from smallholder rope-tied barns to large-scale, free-roaming barns with sleeping cubicles and improved ventilation systems.

This resulted in high production rates and increased export of dairy products. The Netherlands became a famous dairy producer, with the high-producing Holstein Friesian cow as their flagship. Over time, this 'maximization of productivity model' was only adjusted, although the context totally changed. In the process, farmers' knowledge and experience was no longer taken into account, and farmers increasingly depended on subsidies to sustain their income.

Present day dilemmas and problems

First side-effects: environment and surpluses

Since the 1970s, the side-effects of this strategy have become clear: environmental pollution, over-production, dependency on subsidies, decreasing soil fertility, low farmer income and loss of family farms.

The low price and high status of mineral fertilizers made cow manure lose its importance, and was used only as an extra,

---

Box 12.3. Conducive policies in agriculture in the 1950s and 1960s.

The technological gain in Dutch agricultural productivity was heavily supported by a conducive policy environment, which included the following elements:

- Market protection: fixed prices for agricultural products;
- Gaining land by making more 'polders';
- Enlarging existing plots of land for mechanization through 'farmer land exchange' (ruilverkaveling);
- Easy access to credit for farmers (farmer cooperative bank RABO);
- Support to interactive farmer education, extension and research (OVO drieluik);
- Rigorous disease control programmes;
- AI and effective breeding policies;
- Support to agribusiness, leading to low prices of fertilizers and other chemicals;
- Obligation to collect the milk in milk tanks rather than in smaller containers.

Fig. 12.9. Dairy technology development aiming at highest milk yields per animal per year included a shift from hay making from grass to silage making from grass and maize, in order to feed the cattle during the winter period.

over and above the recommended fertilizer application. In 1985, the average fertilizer application for pastures had gone up to 400 kg of nitrogen (N) and 78 kg of phosphorus (P) per hectare. This high fertilizer application, together with high N (proteins) in concentrate feeds, resulted in serious environmental problems because of the leakage of N and P into groundwater and other water bodies. The emission of ammonia (NH$_3$) led to bad smells, contributed to acid rain and affected the quality of nature in the surroundings of the farm.

Meanwhile, at that point, the goal of producing large amounts of low-cost food had already been satisfied. Gradually, another problem arose: the high dairy production levels led to large surpluses of milk powder and butterfat – the so-called milk lakes and butter mountains – which could only be put on the world market with large European subsidies. This led to increasing protests in society in general, as the public concerns for food security had shifted to concern about

environment and international relations. The government was blamed for supporting environmentally polluting production methods, while the subsidized exports were affecting dairy farming in other countries.

Restrictive policy measures

Starting in 1984, the Ministry of Agriculture had to introduce a series of 'restrictive' measures for dairy farmers:

- In order to control the dairy surpluses, the fixed milk prices were banned. A milk-quota system was introduced at European level: farmers could only produce the amount of milk for which they had permission. If they produced more, they were punished financially. If they wanted to produce more milk, they had to buy milk quota from another dairy farmer.

- In order to meet the environmental targets set by the EU, several manure control measures were taken. The total amount of N (artificial fertilizer and animal manure) that could be put on each hectare of the land was limited to a maximum of 250 kg N/ha of grassland (170 at EU level). Spreading manure on pastureland was banned; instead it was made compulsory to inject the manure as slurry into the soil during the growing season (Fig. 12.10).

- In the early 1990s, a mineral bookkeeping system for dairy farmers was introduced. Through accounting of mineral input and output at farm gate level, each farmer had to calculate and report the nutrient losses within the farming system.

As a result of these measures, a clear reduction of milk surpluses as well as mineral surpluses have been seen in Europe since 1985. In The Netherlands, the use of N has been cut by 50%, but it is still twice as high as other European countries like Denmark, Germany and Belgium. Also, the N and P input through concentrated feeds has been reduced since 1985, although a slight increase can be seen after 2005.

These restrictive measures on the one hand limited the negative side-effects of the dairy production system as a whole; on the other hand, they also increased the administrative burden of the farmers, subjecting them to an ever increasing set of rules and regulations.

More side-effects

Ideally, inputs (concentrates and fertilizers) are in balance with the outputs (milk and meat) in terms of nutrients. However, the mineral bookkeeping system in conventional dairy farms revealed that significant losses of N and P occurred in the cow and in the soil. This led to low N efficiency levels (<18% at cow level and <30% at soil level).

Fig. 12.10. Injecting manure as slurry into the soil became obligatory by 1984. The common method of spreading the manure on the fields was banned, in order to reduce the ammonia. The heavy machinery has affected the soil, however, causing soil compaction.

Table 12.2. Technical details of dairy farming in The Netherlands between 1960 and 2007 (Ham *et al.*, 2010).

	1960	1975	1985	1995	2001	2005	2007
Concentrated feeds (kg/cow/year)	830	1890	2280	2210	2000	2020	2120
N with feed (kg N/ha/year)	25	141*	163	182	141	119	156
P with feed (kg P/ha/year)	10	82*	90	70	58	49	61
N in artificial fertilizer (kg N/ha/year)	115	275	350	258	148	146	127
P in artificial fertilizer (kg P/ha/year)	30	30	37	28	17	21	7
Soil surplus N (kg N/ha/year)		350*	400	338	197	184	179
Soil surplus P (kg P/ha/year)		65*	82	59	35	36	15

*In 1980.

The high-input farming system led not only to environmental pollution but also to animal diseases and increasing veterinary costs. Because of the focus on maximization of milk production per animal per year, the animals are pushed to producing high quantities of milk, often at the expense of their health. High protein levels in the feed rations cause digestion problems and malfunctioning of the liver, leading to high incidence of mastitis, hoof diseases, prolonged calving intervals and other fertility problems. As a result, the life expectancy of a dairy cow in The Netherlands is around 4.5 years, which means that she will last no longer than two to three lactations. This not only causes distress in the animal but also in the farmer and his family alike.

Because of the low farm efficiency, the input costs increased, while milk prices were no longer guaranteed. As a result, farmers' income declined. Many of them decided to stop farming or were forced to do so, because of the lack of replacement, as their children did not want to take over the farm. As indicated in Table 12.1, the number of dairy farms in The Netherlands has decreased by 85% between 1960 and 2007. This process is still continuing today.

The main reason for stopping today is related to low milk prices, whereas the inputs are increasing in costs and regulations require major investments. This will only become worse in years to come, as the milk quota system will be abolished by 2015. This will stop the EU-financed cheap exports of dairy products, which has hampered farmers worldwide. On the other hand, it will force many

EU farmers to stop their enterprise, as they will not be able to earn a decent income.

Increasing consumer concern

Since the 1970s, the public in the country has begun challenging the so-called 'licence to produce' of farmers. The first issues were related to the environmental effects of the production methods, as well as the effects of the subsidized exports of dairy products on dairy farming in other (poor) countries. The various outbreaks of infectious animal disease and their control measures, such as the foot-and-mouth outbreak of 2001–2002, resulted in severe crises in society. This aggravated the feeling of discontent amongst the public with large-scale animal production and further exposed the vulnerability of the modern livestock production systems.

Other consumer concerns are related to the effects of cattle on climate change, and animal-related human health issues, like for example the mad cow disease (BSE) that can lead to neurological disorders in humans, and the threat of microbe resistance in humans because of excessive antibiotic use in intensive animal production. More recently, the concerns of the public are especially related to the lack of animal welfare related to intensive production. Dutch people like to see cows in the field when passing through the countryside. The plans for building extra large barns (so-called mega-barns) in which large groups of animals are kept in intensive farms, though potentially with better circumstances in terms of animal well-being, is being confronted with fierce resistance in society today.

In all, Dutch dairy farmers own large farms and produce lots of milk, but that does not necessarily imply that they are rich and can live without worries. On the contrary, many of them are facing serious difficulties, which are often difficult to overcome, while at the same time being criticized by the public (Box 12.4). Fortunately, numerous new initiatives provide important alternatives, which provide ways forward. This will be described in the next part of this study.

Four options and ways forward

In general, one can say that today the options for farmers in response to the problems and dilemmas described above are in the following four directions (van't Hooft, 2010):

Option 1: Stop farming altogether or start a dairy farm abroad

Every week, around 50 farms in The Netherlands decide to stop their activities, mostly related to lack of economic profitability or inability to find a replacement to run them. Some farmers try to find other employment activities; others decide to start a dairy farm in another country, often in Eastern Europe, where land is cheap and few regulations limit the dairy enterprise.

In most cases, the land and the milk quota of the original farm is bought by a neighbouring farmer, in order to go for option 2: scale enlargement.

Option 2: Further scale enlargement and technical efficiency

Another section of the farmers opt for scale enlargement and technical efficiency of their farming business. New technical developments include the milking robot, automatic feeding and manure cleaning devices for the barn (Fig. 12.11), which also include devices for filtering the air for greenhouse gases. New barn concepts (so-called mega-barns) include animal well-being aspects for 250 to up to 1000 cattle, with enough space as if they were roaming free (Fig. 12.12). Some concepts even go as far as developing complete agro-production parks, in which livestock keeping is totally confined within an industrial area, without the link to the traditional landscape.

The scale enlargement option is supported by mainstream science, as well as ministries and major formal farmer organizations. At the same time, it is encountering increasing resistance from the public, amongst other things for landscape and cultural reasons. If this scale enlargement option does not solve the other basic problems of the conventional dairy farming model, such as the loss of soil fertility and dependency on chemical inputs and

---

Box 12.4. Dilemmas and problems of modern dairy farming in The Netherlands.

The dilemmas:

- Super specialization and productivity focus;
- Decreasing soil fertility;
- Dependence on inputs (fertilizer, carbon energy) and subsidies;
- Environmental problems;
- Regulation requires administrative burden and major investments;
- Need to reduce antibiotic use – but how?

Resulting in:

- Low income because of very low profit rate per kg of milk produced;
- Income prospects are difficult because of milk quota system being abolished in 2015;
- Social problems – farmers going out of business (85% have stopped between 1960 and 2007);
- Young people moving out of farming, lack of 'replacement';
- Criticism of public, especially on animal well-being and climate change.

Fig. 12.11. New technical developments include the milking robot and (in this picture) the manure cleaning robot.

Fig. 12.12. Some new barn types are designed as if the animals are in the fields.

carbon (C) energy, in time it may well run into the same economic problems that smaller farms are facing today.

Option 3: Diversification of farm income through multi-functional farms

Recent research (Oostindie *et al.*, 2011) has shown that around 40% of the Dutch farms today have one or several income-generating activities besides the primary productive activity to add to the farm income. These may include activities directly linked to farming as well as non-farming activities linked to the public.

Second sources of income directly linked to farming:

- Making cheese, yoghurt or other dairy product;
- Service to nature and environment, such as wild birds;
- Second productive activity, such as organic chicken;
- Bio-digester;
- Energy generation through solar panels on barn roof.

Non-farming activities linked to the public:

- Shop with organic farm produce or regional products;
- Care farms for specific groups, such as elderly, physically or mentally handicapped, youngsters, people with burn-out etc.;
- Camping site, meeting facilities, bed-and-breakfast;
- Children's parties, educational activities for schools;
- Nature walks through farming premises.

These non-farming activities are linked to the tendency amongst urban people to establish linkages with people and farms in rural areas. They are especially viable in rural areas close to the urban centres (Figs 12.13 and 12.14).

Option 4: Increasing income through increased mineral efficiency, soil fertility and cost reduction

An increasing number of dairy farmers are engaged in improving their income through improved soil fertility and farm efficiency

Fig. 12.13. Farmers' golf has become one of the activities that dairy farmers offer to urban citizens. The golf game takes place with special sticks and balls in the dairy fields amidst the animals.

Fig. 12.14. Urban citizens increasingly recognize the need to show their children from where the milk and other food comes. This may explain the success of the children's parties on the dairy farms.

(Fig. 12.15). They aim to 'go with nature' and build on natural processes, rather than go against nature and control the natural processes. They aim to close the nutrient (N, P, C) cycles and footprint at farm, region and international level. This option has two major groups of farmers: certified organic farmers and conventional farmers that improve their farm efficiency through the 'soil–plant–animal–manure approach' (van Weperen and Kieft, 2002).

Organic farming is the oldest form of this option; biodynamic farming adds a spiritual element to organic farming. These organic farmers receive a special price for their products based on an organic/biodynamic certification. The products are sold in special shops and increasingly also in supermarkets. Though the market for organic produce is increasing, the number of organic dairy farms has remained very limited (1.3% of total). This is very low compared with a number of other European countries: Austria 12.8%, Denmark 9.6%, Sweden 6.6% and Switzerland 5.5% (www.biokennis.nl).

At the same time, an increasing number of conventional farmers are working with the 'soil–plant–animal–manure approach' towards reduced inputs of fertilizer and concentrated feeds, through increased farm efficiency and soil fertility. In so-called study groups (Proost and van Weperen, 2006), they are learning to:

• Apply less artificial fertilizer on fields, adapted to region and soil type;

• Produce milk on basis of roughage rather than concentrates: feed the cow as ruminant again;

• Use management practices that enhance soil life and soil organic matter, closely linked to the quality of animal manure;

• Monitor the link between mineral efficiency (especially N and P) and farm income;

• Establish effective combination with nature preservation (e.g. wild birds) and receive some extra income from this.

Fig. 12.15. Farmers increasingly see the potential of improving their soil fertility in order to increase their financial returns. It has become popular to learn from each other in study groups, to visit each other's farm and compare monitoring data that result from their dairy enterprise.

Nine traditional elements in postmodern sustainable dairy farming

One could say that the options 3 and 4 represent the initiatives that were started by family farms themselves in order to become more sustainable in the long run. Not only are these initiatives initiated by farmers themselves, they also include several traditional elements from the past that are now being re-discovered in order to find appropriate ways to address the present-day problems. For that reason, they may well be called the ways towards 'postmodern, sustainable family farming'.

Traditional elements of livestock keeping (also known as ethno-vet) are thus playing a new role in Dutch dairy farming. One could indentify eight ways in which this is taking place.

Element 1: Build on farmers' initiatives and knowledge

Instead of blanket recommendations, it is now time to make room for farmer initiatives and knowledge. The successful (dairy) farming families of today have found innovative ways out of the dilemmas they are facing, building on their own traditions, local circumstances and innovative ideas. Many of these initiatives include practical ways to 'go with nature' instead of against it.

Research and support organizations increasingly build on this and support the farmers in this quest, joining them in the experiments and providing answers to the new questions these farmers are facing.

Presently, this is being done in numerous 'farmer learning networks' that are supported by the government as well as private organizations.

Element 2: Re-establish the importance of the soil–plant–animal–manure cycle

The focus on the farm as a natural whole, with its soil–plant–animal–manure cycles, is basic in the sustainable approach to dairy farming. The aim is to close the N, P and C nutrient cycles in order to reduce costs, increase farm efficiency and farm income. This cycle was broken when dairy farming specialized and became separate from crop farming and the production of a major part of the animal feed.

This soil–plant–animal–manure approach was originally started by a group of concerned farmers in the northern province of Friesland in the early 1990s. The philosophy of these dairy farmers was based on their 'gut feeling' and experience with the production system of their fathers. They realized that many relationships exist between the various components of the farming system, and that these relationships should be considered in all management decisions. They transformed this into practical management recommendations, which are adapted to soil type and other local circumstances. In spite of limited support from research and policies, this concept has gradually grown amongst dairy farmers in the country. Groups of dairy farmers are especially encouraged by the project Duurzaam Boer Blijven, which roughly translates as 'Continue Farming the Sustainable Way'.

Element 3: Focus on farm optimization rather than maximization of a single product

This focus on the soil–plant–animal–manure cycle also implies a shift away from the focus on the maximization of a single commodity product, such as milk. Rather, it aims to find an economically and ecologically optimal balance between different elements of the system. In that way, the amount of milk produced per animal per year is not the major parameter for progress (Stuiver *et al.*, 2003; Verhoeven *et al.*, 2003).

One of the ways this is translated is the aim for the highest 'life production' rather than 'year-production'. In this way, it is more economical to maintain animals during a longer life period. This has advantages in many ways, including an effective reduction of the greenhouse gas emissions, because fewer replacement animals are needed.

In The Netherlands, the farmer with so-called 100,000-l cows is now rewarded: a cow that has produced 100,000 l throughout her life. These cows are of excellent health and productivity, as they have had at least 10 lactations with excellent results (Fig. 12.16). Most of these cows do not have their highest production rates during their first lactations; they usually produce most

during their sixth or seventh lactation. This is another reason for adapting the present system, in which the average cow only reaches 4.5 years of age.

Element 4: Re-diversifying the farmer's work and income

As indicated above, around 40% of the Dutch farmers are gaining an extra income from secondary sources, either with activities directly linked with the dairy farm, or indirectly through activities that involve the public. Increasingly, specific arrangements are developed that link urban people with rural farms over a longer period, like for example:

- Open days when the public can visit farms;
- Consumers can adopt a cow, chicken or apple tree, and become a 'friend' of a certain farm;
- Arrangements where the public can buy 'shares' for which farm produce can be purchased.

In many farms, a number of these activities are combined. Sometimes these secondary activities have become economically far more important than the original (dairy) production on the farm. Though originally a farmer-driven initiative, this option is now formally supported by ministries and farmer organizations.

Fig. 12.16. A dairy farmer is especially proud of a 100,000-l cow. This implies a sustainable management system in which a cow can live and produce at excellent levels during at least 10–11 lactations. The focus on high 'life production' is more sustainable than the focus on highest milk yield per animal per year.

Element 5: Re-valuing direct marketing and urban–rural linkages

Family farming is increasingly seen as a cultural asset within Dutch a society. If only very few large-scale farming enterprises were to remain in the country, we would lose a major part of our typical landscape and cultural background. That is one of the reasons for the success of the rural–urban linkages today that provide secondary sources of income.

One of the problems here is that urban consumers do not always translate this concern into buying organic products in the supermarket. The marketing of organic produce is growing but cannot keep up with the speed of the loss of family farms. Shortening the market chains through direct linkages between urban consumers and rural producers is one of the key elements that keep family farms from disappearing, while at the same time stimulating urban consumers not only to go for the cheapest option in the supermarket.

Element 6: Re-valuing the link between farming and the natural environment

The increased recognition that farming has to go hand-in-hand with nature stimulated farmers to integrate other natural elements into their faming system (Fig. 12.17). For example, the farmers organized in the so-called Agrarische Natuur Verenigingen, roughly translated as 'Farmer Groups that Promote Nature', receive an EU subsidy for adapted field management to maintain wild birds. They have a contract with the government that they will not cut the fodder in certain fields before 15 June, providing space for wild birds to reproduce.

The development of these so-called 'green and blue services' are still quite premature. Research has shown that these nature conservation and landscape management practices have improved the incomes of the farming families: around 10% of their income is generated by means of nature programmes financed by the EU and the Dutch national government (de Rooy, 2010).

Element 7: Re-valuing local and dual-purpose breeds

There is a limited but increasing tendency to re-value Dutch local cattle breeds (as well as other species), such as Lakenvelder, Blaarkop, Brandrode runderen, as well as dual-purpose/beef animals from other EU countries, like Belgium Blue and Vleckvieh. Sometimes these are used as grazers in natural areas, but increasingly also in dairy farms.

The reasons include:

- These dual-purpose animals are more efficient in digesting roughage than the pure milk breeds.
- These animals provide income from two commodities: milk and meat. For example, meat calves fetch a better

Fig. 12.17. Dairy farms are often located near natural areas. So-called 'green and blue services' can provide an extra income for dairy farmers, when their management practices promote wildlife, soil and water quality.

price than milk calves; after milking, cows are fattened more easily for sale.

- These animals are more robust and are better equipped for good health and long life. In spite of lower annual production than the traditional dairy breeds, they often end up with a higher total life production.
- These more robust animals have fewer health problems than the pure milk breeds, which means fewer troubles for the farmer (and his family).

Element 8: Re-linkage between crop and livestock farms

Most Dutch dairy farms are not integrating food crops and livestock. They are specialized farms in which the manure of their animals is used for their pastures and other fodder crops, but they are limited by law to do so. In some cases, (organic) dairy farmers make strategic alliances with crop farmers nearby, in which surplus manure is used to fertilize the food crops, whereas the crop residues are used as feedstuff for the cattle. In this way, the traditional linkages between crops and animals are re-established. It is expected that this tendency will grow in the near future.

Element 9: Re-valuing medicinal plants

The main antibiotic use in dairy is related to udder health, for curing and prevention of udder infections. In general in The Netherlands, the use of chemicals and antibiotics in specialized livestock farming is high, much higher than in neighbouring countries.

This is of increasing concern, especially because of the problems related to resistant microbes in pig production of MRSA (multi-resistant *Staphylococcus aureus*) and in poultry production of ESBL (extended spectrum B lactamase in microbes – see www.RIVM.nl). Both of these microbes pose serious threats to human health, as people infected with these microbes will not recover from an infection by using antibiotics. Therefore, the government has decided that the use of antibiotics in Dutch livestock farming has to be reduced by 50% before 2014.

Meanwhile, most of the knowledge of our ancestors related to the use of medicinal plants in this country has been lost. A study by the Institute for Ethnobotany and Zoopharmacognosy (IEZ) revealed 168 folk remedies with 68 plant genera still exist within The Netherlands. Similar results come from studies in other EU countries (van Asseldonk and Beijer, 2006).

Medicinal plants are most frequently used in organic dairy farming. There are presently around 255 commercial products for farm animals, most of them as feed additives in a reaction to the EU ban on antibiotic growth promoters in 2006. Research in quality, efficacy and safety is lacking and formal backing from Veterinary University is missing. The use of medicinal plants is banned from the veterinary curriculum. Meanwhile, new initiatives related to the use of medicinal plants in livestock keeping are being developed throughout the country, and a course on herbal medicine for animals is now being offered. The potential of herbal animal remedies is slowly dawning.

Main conclusions

When intensification of the dairy production in The Netherlands started in the 1960s it was not clear what the side-effects of this exercise would be. Today, the history of Dutch modern dairy farming – besides the staggering productivity gains – clearly demonstrates the side-effects of the productivity paradigm in terms of social, environmental, economic and animal well-being consequences. What would happen in a developing country, for example, if a majority of the smallholder farmers were to stop farming and move to the urban centres?

The agricultural development within The Netherlands has been based on super-specialization in only milk, meat or crop production of one single species, aiming at urban markets and exports. The spectacular productivity increase was largely related to the conducive policy environment between 1950 and 1980. It is not easy to reproduce this process without similar market interventions and other conducive policy measures. Meanwhile, times have changed, and under the present international financial politics, these policy measures are very difficult to reproduce in the developing country context.

In The Netherlands today, a contradiction remains. Most mainstream research and policies aim at further scale enlargement and technological development, while new farmers' initiatives frequently aim at diversifying farm income and optimizing rather than maximizing farm productivity. These innovative farmers with their integrated knowledge and wide networks can be very helpful for dissemination, research and policy towards sustainable dairy development; in fact, they stand at the basis of agricultural transition.

At the same time, herbal medicine – like other traditional techniques – has been lost to a large extent in The Netherlands and other EU countries. Perhaps we could build on experiences from developing countries in its recovery – especially with the aim of reducing antibiotic use?

In turn, the developing countries, rather than copying the intensive dairy production system, could learn from the experience in The Netherlands to prevent loss of family farms, biodiversity, rural–urban linkages, integrated farming and related knowledge. The postmodern sustainable dairy farming initiatives in The Netherlands today are recovering nine different elements of traditional agriculture, that can still be found in most developing countries.

Therefore, developing countries would be wise to develop their own strategy starting with their own resources and local circumstances, rather than copying the dairy system from The Netherlands (Fig. 12.18). In this way, the developing countries can make a 'technology leap' and develop their own sustainable dairy systems, while preventing the negative side-effects of modern dairy farming that are now known to us.

Fig. 12.18. Dairy initiatives in developing countries do not need to copy the Dutch dairy development model. Instead they can make a 'technology leap' and prevent the side-effects of intensive dairy farming that have become apparent in The Netherlands.

Relevant websites

www.dutchfarmexperience.com
www.duurzaamboerblijven.nl
www.ethnobotany.nl

www.verantwoordeveehouderij.nl
www.multifunctionelelandbouw.nl
www.biokennis.nl
www.rivm.nl

References and Further Information

Asseldonk, A.G.M. van and Beijer, H. (2006) Herbal folk remedies for animal health in The Netherlands. Paper for International Ethnobotany Conference.

Ham, A. van den, Berkmortel, N. van der, Reijs, J., Doornewaard, G., Hoogendam, K. and Daatselaar, C. (2010) *Mineralenmanagemet en economie op melkveebedrijven. Gegevens uit de praktijk* (*Mineral Management and Economics of Dairy Farms. Data from Practice*). Brochure 09-066, February. LEI, Wageningen UR.

Hooft, K. van't (2010) The role of ethnoveterinary practices in post-modern agriculture: examples from The Netherlands. In: *Ethno-Veterinary Practices. Mainstreaming traditional wisdom on livestock keeping and herbal medicine for sustainable rural livelihood. Proceedings of the International Conference on Ethno-Veterinary Practices,* jointly organized by TANUVAS and I-AIM at South Zone Cultural Centre, Thanjavur, Tamil Nadu, 4–6 January.

Oostindie, H., Seuneke, P., Broekhuizen, R. van, Heggen, E. and Wiskerke, H. (2011) Dynamiek en robuust-heid van multifuncionele landbouw (Dynamic and strength of Multifunctional Agriculture). Report phase 2, WUR, January.

Proost, J. and Weperen, W. van (2006) Creating space for change: farmers' learning groups in The Netherlands. In: *Compas Magazine*, no. 10, July.

Rooy, S. de (2010) The Netherlands – The VEL/VANLA environmental co-operatives as examples of endog-enous rural development through farming. In: *Endogenous Development in Europe*, Chapter 2. Compas, Leusden.

Stuiver, M., Ploeg, J.D. van der and Leeuwis, C. (2003) The VEL and VANLA environmental co-operatives as fields laboratories. In: *Rethinking environmental management in Dutch dairy farming: a multidisciplinary farmer-driven approach. NJAS, Wageningen Journal of Life Sciences* 51, 27–39.

Verhoeven, F.P.M, Reijs, J.W. and Ploeg, J.D. van der (2003) Re-balancing soil-plant-animal interactions: towards reduction of nitrogen losses. In: *Rethinking environmental management in Dutch dairy farming: a multidisciplinary farmer-driven approach. NJAS, Wageningen Journal of Life Sciences* 51, 147–164.

Weperen, W. van and Kieft, H. (2002) Dutch dairy farmers find own solutions to their environmental problems. In: *LEISA Magazine*, April.

Appendix

Overview of Recommendations for Optimizing Smallholder Livestock Keeping

(1) ANIMAL NUTRITION	Objectives	Dry season nutrition	Mineral supply
Low-input and diversified smallholder systems	Reduced mortality in dry season Reduced weight-loss Increased resistance to drought		
Recommendations for improvement		Agricultural left-overs storage and feeding Support local feeding innovations Plant leguminous trees Improved use of kitchen left-overs Green forage Hay making Cheap by-products Feeder troughs	Provide ordinary salt Home-made mineral blocks Vitamins
More specialized smallholder systems	Better nutritional status year-round Improved reproduction rate Special feeding young stock		
Recommendations for improvement		Local production of balanced feed Improved straw feeding Hay/silage making	Complete mineral supplements Vitamins

Continued

©K.E. van 't Hooft, T.S. Wollen and D.P. Bhandari 2012. *Sustainable Livestock Management for Poverty Alleviation and Food Security* (K. van 't Hooft, T. Wollen and D.P. Bhandari)

Continued.

(2) PASTURE AND RANGELANDS	Objectives	Pasture	Rangeland
Low-input and diversified smallholder systems	Reduced overgrazing and soil-erosion Reduced bush encroachment Increased carrying capacity Increased resistance against drought Community organization		
Recommendations for improvement		Controlled grazing Zero-grazing system	Reviving communal grazing control Fencing off grazing areas Rotational grazing Special grazing areas for dry period Controlled and prescribed fire
More specialized smallholder systems	Sufficient fodder available year round Good quality fodder Good N and P efficiency Increase soil fertility and soil life		
Recommendations for improvement		Plant fodder crops Pasture rotation Special pastures for young stock Zero-grazing system Efficient fertilization of pastures	Effective weed control
(3) WATER	*Objectives*	Access to water	Water quality
Low-input and diversified smallholder systems	Regular water uptake Water quality sufficient Pollution of human water sources prevented		
Recommendations for improvement		Water 1–2 times a day Opt for animal species that require little water	Prevent polluted drinking water for animals Prevent pollution of water for human use by animals
More specialized smallholder systems	Good water availability and year round quality		
Recommendations for improvement		Continuous access or provide access 3–4 times a day	Prevent pollution of drinking water by chemicals or artificial fertilizers

Continued

Continued.

(4) INFECTIOUS DISEASES	Objectives	Animal health services	Vaccination
Low-input and diversified smallholder systems	Reduced incidence zoonosis Reduced animal mortality due to infectious disease Promote synergy between traditional and modern remedies Improved access to local animal health services		
Recommendations for improvement		Support ethno-vet practices and practitioners Train Community Animal Health workers (CAHW) Awareness about zoonosis	Vaccination of one or two against major infectious diseases
More specialized smallholder systems	Improved use of ethno-vet medicine (medicinal plants) Improved use of commercial medicine One health: human and animal medicine join forces		
Recommendations for improvement		Ethno-vet practices strengthened Training improved use of commercial medicine Disease surveillance Monitoring and recording of disease incidence	Extended vaccination programmes

(5) PARASITE CONTROL	Objectives	Internal parasites	External parasites
Low-input and diversified smallholder systems	Reduced incidence of internal and external parasites Prevention parasitic zoonosis Reduced loss young stock due to parasites Improved leather quality		

Continued

Continued.

(5) PARASITE CONTROL	*Objectives*	Internal parasites	External parasites
Recommendations for improvement		Make use of natural resistance of local breeds Reducing parasite incidence in grazing and feeding areas Parasite control especially in young stock Support ethno vet remedies for parasite control	Make use of natural resistance of local breeds Use of medicinal plants for parasite control (ethno vet) Community control activities (bathing)
More specialized smallholder systems	Low incidence of internal and external parasites in all stock Special care for young stock (especially exotics and crossbreeds) Prevent resistance against commercial medication		
Recommendations for improvement		Regular treatment all stock Medicinal plants and commercial medications	Regular bathing/ spraying of all stock Medicinal plants and commercial medications

(6) BREEDING	*Objectives*	Use of breeds	Breeding management
Low-input and diversified smallholder systems	Maintain important local breeds Make use of important traits of local breeds Effective selection Prevent inbreeding		
Recommendations for improvement		Breeding selection on basis of local criteria Bring in improved local breeds	Prevent inbreeding Timely castration Change males before mating with own offspring
More specialized smallholder systems	Increased productivity Effective selection Selective use of exotics Good reproductions rates Maintain local breeds for crossbreeding		

Continued

Continued.

(7) PROTECTION AND HOUSING	Objectives	Predators, accidents and theft prevention	Weather protection
Recommendations for improvement		Improved local breeds Crossbreeding between 25–75% of exotic blood Selection of bulls Selective/limited use of artificial insemination	Breeding only at minimum age and height Strict control of uterus infection Effective heat detection (in case of AI)
Low-input and diversified smallholder systems	Reduced loss due to predators, theft and trampling Effective low-cost constructions with local materials Prevent transmission of zoonotic parasites		
Recommendations for improvement		Protection of young animals during first weeks Protection during brooding and caring for young Night shelters No contact between animals and human stool	Provide simple night shelters Trees for shade in fields
More specialized smallholder systems	Effective housing with enough space for each animal Clean housing Manure available for crops		
Recommendations for improvement		Milking shed Ventilation Manure-pit	Housing for zero-grazing
(8) SPECIAL CARE	Objectives	Sick animals	Around delivery
Low-input and diversified smallholder systems	Increased survival rate of sick animals Reduced mortality of newly born Reduced disease female animals after birth Good bonding Reduced disease transmission around birth		

Continued

Continued.

(8) SPECIAL CARE	Objectives	Sick animals	Around delivery
Recommendations for improvement		Separate sick from healthy animals Shade, water, fresh food Ethno vet treatment Disposal of dead animals	Have animal near before birth Attend birth when necessary Check afterbirth Guarantee colostrum intake
More specialized smallholder systems	Effective control of high-productivity problems Young stock in good condition		
Recommendations for improvement		Special care of young stock Regular mastitis control	Support to difficult birth due to large fetus
More specialized systems		Ethno vet/commercial treatment	Milk fever control

(9) IMPROVED MARKETING	Objectives	Informal Markets	Formal – regional markets
Low-input and diversified smallholder systems	To supply family and local community requirements for food and household needs To fully utilize local resources	Barter and other forms of exchange Sale at the doorstep	Develop niche markets of local breeds
Recommendations for improvement		Support local innovation in production, storage and marketing Improve product quality and volume Develop niche markets Improved organization	Improve product quality and volume Improve communication about market opportunities
More specialized smallholder systems	To maximize production volume, uniform quality To meet consumer desires		
Recommendations for improvement			Improve product quality and volume Develop niche markets for organic or environmentally sound products Develop sell and buy cooperatives that provide uniform delivery through high and low seasons

Index